Nuclear Economy 2

SCIENCES

Energy, Field Directors – Alain Dollet and Pascal Brault

Nuclear Energy, Subject Head – Sylvain David

Nuclear Economy 2

Nuclear Issues in the Energy Transition

Coordinated by
Jacques Percebois
Nicolas Thiollière

ISTE

WILEY

First published 2023 in Great Britain and the United States by ISTE Ltd and John Wiley & Sons, Inc.

ISTE Ltd
27-37 St George's Road
London SW19 4EU
UK

www.iste.co.uk

John Wiley & Sons, Inc.
111 River Street
Hoboken, NJ 07030
USA

www.wiley.com

Library of Congress Control Number: 2023931851

British Library Cataloguing-in-Publication Data
A CIP record for this book is available from the British Library
ISBN 978-1-78945-095-8

ERC code:
PE2 Fundamental Constituents of Matter
PE2_3 Nuclear physics
PE8 Products and Processes Engineering
PE8_6 Energy processes engineering
SH1 Individuals, Markets and Organisations
SH1_12 Agricultural economics; energy economics; environmental economics

Contents

1

Management of Spent Fuels

Jacques CANTARELLA[1], Cécile EVANS[2], Pierre KUNSCH[1,3], Didier LÉONARD[1] and Jean-Paul MINON[1]

[1]*ONDRAF, Brussels, Belgium*

[2]*Orano, Châtillon, France*

[3]*École polytechnique de Bruxelles, Brussels, Belgium*

1.1. Spent fuel management strategies and costs

One of the main issues influencing the decisions made regarding the use of nuclear energy concerns the management of spent fuels, containing both highly radioactive waste and recoverable nuclear materials (namely fissile uranium and plutonium). The spent fuel management technologies that are used must meet the following requirements: guaranteeing the fundamental safety objective of protecting people and the environment from the dangers of radiation, ensure compliance with guarantees against nuclear proliferation and the safety of nuclear materials, and maintaining the economic aspects of nuclear energy

Since commercial nuclear energy operations first came into existence about 70 years ago, the nuclear sector has taken a responsible approach to the entire lifecycle of spent fuels. For several decades, some countries have been applying the principles of circular economies through the recycling of spent fuel and through minimizing waste. In other countries, the effective implementation of safe surface storage solutions has provided the time needed to develop robust technical solutions, as well as a democratic and transparent decision-making process for the final storage of high radioactivity waste. In any case, the final storage of highly radioactive waste is an issue that requires close attention. There is a widespread belief today, grounded in scientific study, that the removal of high-level waste (HLW) from the human

Nuclear Economy 2,
coordinated by Jacques PERCEBOIS and Nicolas THIOLLIÈRE.

environment through storage in deep geological repositories is both safe and environmentally friendly, and that the science and technology behind this practice are well developed. However, no deep geological repositories are operational in the world at this time.

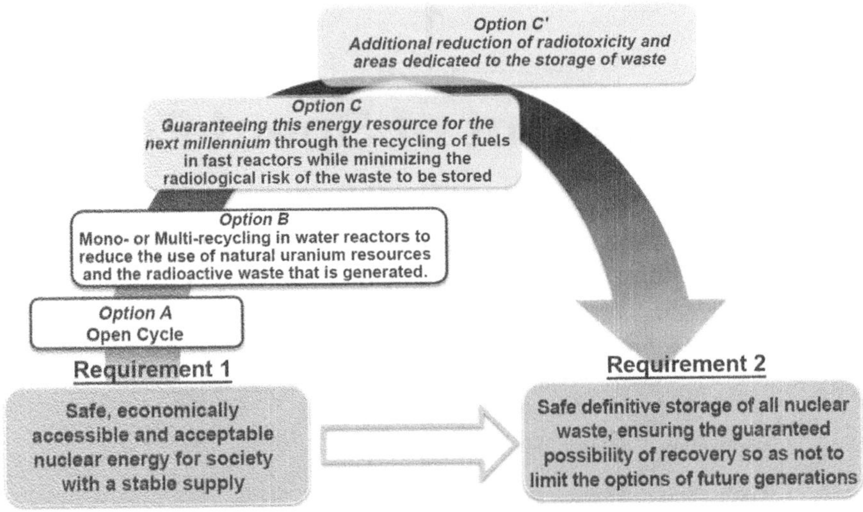

Figure 1.1. *Requirements and large cycle options presented in IAEA (2018). For a color version of this figure, see www.iste.co.uk/percebois/nuclear2.zip*

1.1.1. *The different options of the cycle*

The different options of the cycle offer a different status to the spent fuel:

– the fuel in its entirety is considered to be waste, and as such must be packaged and isolated in deep storage;

– the fuel is recycled in the water reactors in order to use the small amount of remaining fissile material;

– the fuel is recycled several times in Generation IV breeder reactors to convert and use the large remaining amount of fertile materials.

These three primary options show the progress both in terms of the circular economy, but also in terms of minimizing the amount of highly radioactive waste to be managed over the very long-term (OECD/NEA 2006, 2021) as illustrated by Figure 1.1.

1.1.1.1. *The open cycle*

The open cycle (see Figure 1.2) is the fuel cycle currently considered as the reference option in a number of nuclear-powered countries. It is based on thermal neutron reactors.

The spent fuels are stored safely, waiting to be placed in a deep geological repository. The surface storage of the fuels can take place on the reactor site and/or on a dedicated centralized site. The spent fuels will be transported to the packaging site in waste packages compatible with the requirements of the deep storage infrastructure. The spent fuels do not undergo chemical treatment since they are considered to be fully composed of highly radioactive waste.

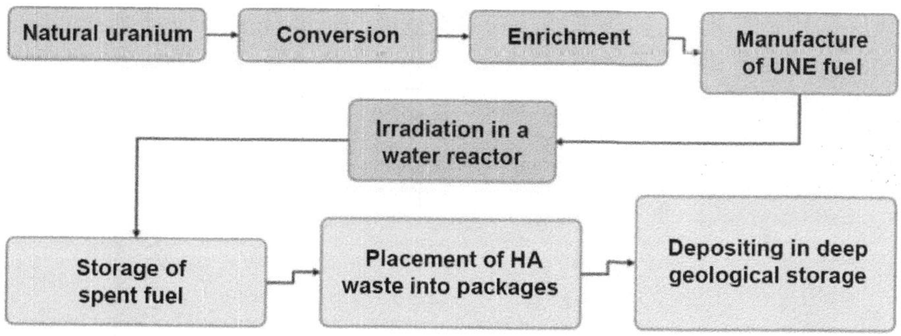

Figure 1.2. *Illustration of an open nuclear cycle. For a color version of this figure, see www.iste.co.uk/percebois/nuclear2.zip*

1.1.1.2. *Monorecycling in a light water reactor*

The process of monorecycling in a light water reactor (see Figure 1.3) has been implemented for several decades in France, as well as in a number of other European countries, Russia and Japan, and has now gained benefits from significant industrial experience. It is currently being implemented in China with the planned development of recycling capacities. The recycling facilities of Orano de La Hague and Melox have processed more than 37,500 tHM (metric tons of heavy metals) of spent fuels discharged from French and foreign reactors, and manufactured 2,900 tHM of MOX fuels loaded in more than 40 light water reactors, and are considered to be state of the art at the global level. This option is the first step to progressing toward more advanced options for the multi-recycling of fuels in light water reactors through the use of fuel technologies in development of the MOX2 or Remix type, and/or toward the closure of the cycle through the use of fast neutron reactors.

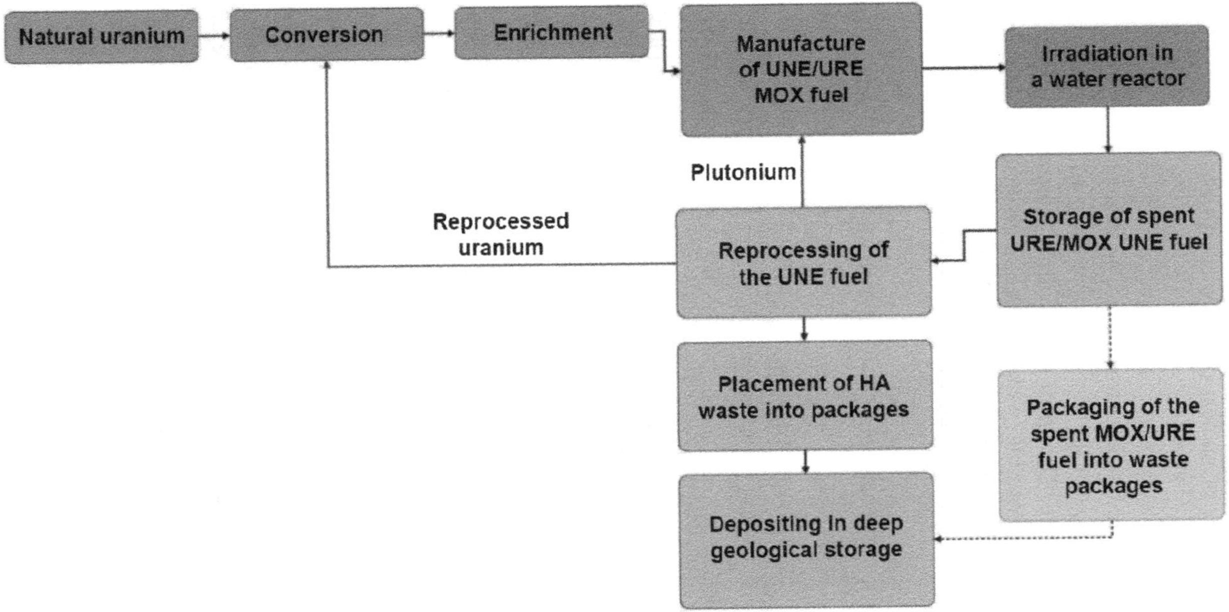

Figure 1.3. *Illustration of a nuclear cycle with mono-recycling in a light water reactor. For a color version of this figure, see www.iste.co.uk/percebois/nuclear2.zip*

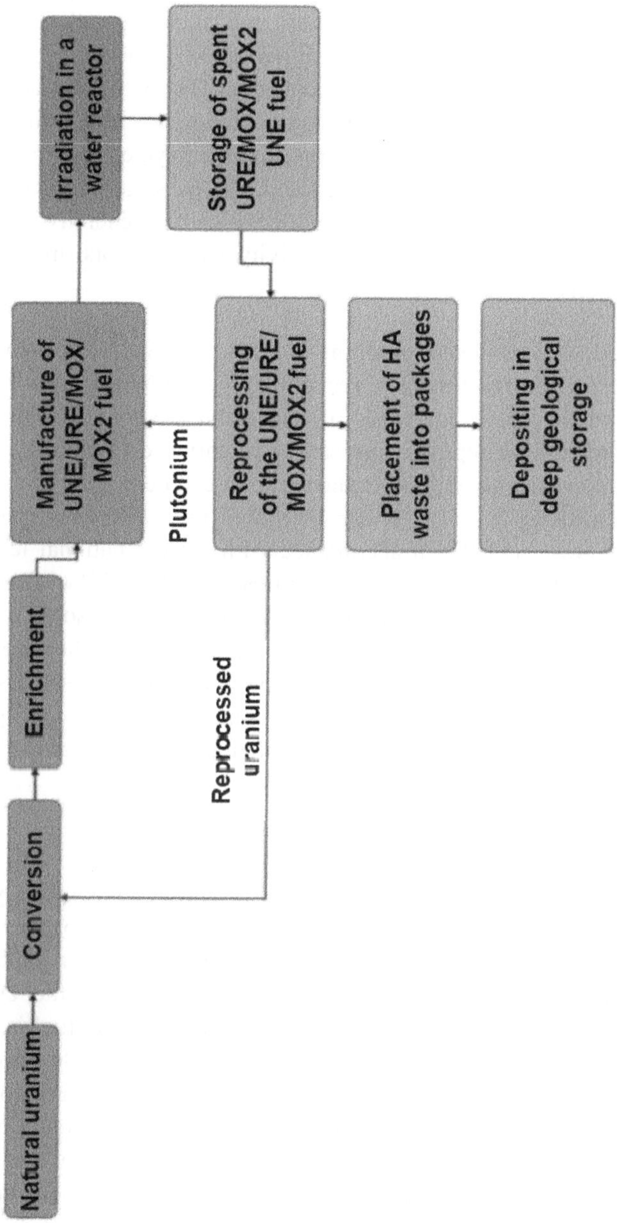

Figure 1.4. *Illustration of a nuclear cycle with multi-recycling in a light water reactor. For a color version of this figure, see www.iste.co.uk/percebois/nuclear2.zip*

Figure 1.3 illustrates the processes involved. The cooled spent fuel is chemically separated (reprocessed), and the uranium and the plutonium for reprocessing are recovered to be reused in the form of new fuels; the minor actinides and fission products constituting the "high-level waste" are packaged in the form of a glazed, solid and stable matrix inside stainless steel containers. The vitrified packages are stored while waiting to be placed in a deep geological repository. The volume of this waste is less than the volume of spent fuel from the open cycle (by a factor of 5). The metal structures of the fuel are compacted and also packaged in stainless steel containers, constituting medium radioactivity waste which can generally be disposed of safely in properly designed medium radioactivity waste depots, in accordance with national regulations.

Plutonium mixed with depleted uranium is used to make mixed oxide fuel (MOX), which is refilled in reactors. The recycling of plutonium makes it possible to use the content of the fissile materials of plutonium (about 60%) in the fuel. The uranium separated during reprocessing can be re-enriched to make a new fuel. The reprocessed uranium is only slightly radioactive; it can be preserved as a future resource. If it is considered a waste, it can be stored in a way that is similar to long-lived low-level waste (LLW) in accordance with current national legislation. This fuel cycle, qualified as mono recycling, requires only about 75% of the natural uranium necessary for the open cycle due to the reuse of the fissile isotopes uranium and plutonium. An evolution toward a multi-recycling cycle in a light water reactor can be projected (see Figure 1.4), pending the deployment of fast reactors. In this sense, France is currently conducting a research and development program on MOX2 fuel technologies with objectives of stabilizing spent fuel and plutonium inventories.

1.1.1.3. *Multi-recycling in a fast neutron reactor*

Multi-recycling in a fast neutron reactor (see Figure 1.5) has not yet been implemented on an industrial scale and continues to be the subject of research and development in many countries, as well as within the framework of multinational collaborations. As the name suggests, this fuel cycle is similar to the fuel cycle with multi-recycling in a water reactor described above, except that the fuel is recycled several times in fast neutron reactors. In order to have sufficient fissile materials for very long-term multi-recycling, a fast neutron reactor is used. This type of reactor uses neutrons at higher energies, resulting in an increase in the plutonium generated from the fertile uranium. Approximately the same amount of plutonium is obtained from the spent fuel as in the fresh fuel, or even more plutonium when used in the breeder mode. The minor actinides and the fission products produced during each cycle must be separated and packaged in waste packages, in a similar way to the recycling cycle in a water reactor.

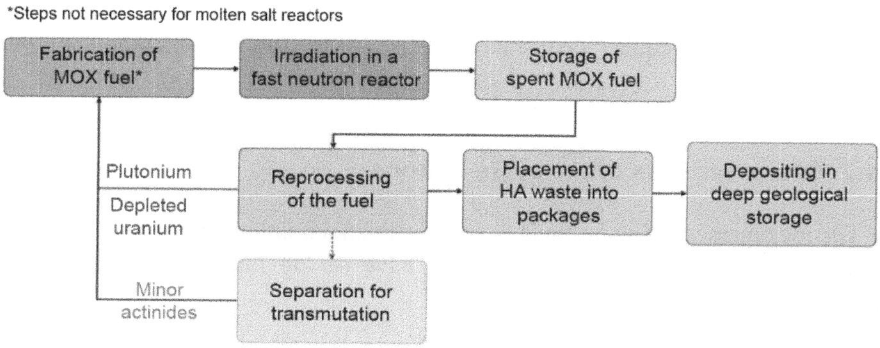

Figure 1.5. *Illustration of a nuclear cycle with multi-recycling in a fast neutron reactor. For a color version of this figure, see www.iste.co.uk/percebois/nuclear2.zip*

In a fast neutron reactor, the conversion of fertile uranium into fissile plutonium is very high in comparison with a thermal reactor. This implies a sharp reduction in the amount of natural uranium necessary for the production of a unit of energy.

Fast reactors can recycle the plutonium they produce or plutonium produced by water reactors indefinitely, including that of MOX or MOX2 light water reactor fuels, thus making the most of their energy potential while avoiding the long-term storage of spent fuel that contains valuable materials.

Advanced options can also transmute minor actinides (in particular neptunium, americium and curium) in order to further reduce the amount of HLW, as well as the duration for which they will remain hazardous. These options are under development and require additional technologies during the flow separation step, and for the manufacturing step of the fuel. Fuels containing minor actinides are highly radioactive and must be manufactured remotely in specially shielded "hot cells". To obtain a sufficient transmutation of minor actinides and potentially long-lived fission products, different systems may be called for, such as sodium fast neutron reactors, molten salt fast neutron reactors, or possibly accelerator-controlled subcritical reactors (referred to as ADS, which stands for accelerator-driven system).

In the case of molten salt reactors, the fuel is liquid, so the requirements of the recycling system are different from other systems operating with solid fuels. We can note a recent revival of interest in research on the transmutation separation with the objective of reducing the amount of radioactive waste, its residual heat and its long-term radiotoxicity, and consequently, the need for deep storage infrastructures is reduced (OECD/NEA 2011). However, this additional step, now in the initial R&D phase, will require dedicated cycle infrastructures and adds additional expenses.

1.1.2. *Characteristics of the cycle options*

The management of spent fuels is a major financial issue that requires long-term strategic planning. Today, most countries have set up financing mechanisms to generate funds for this purpose.

The construction and operation of the geological storage facility represent significant costs that usually occur decades after the energy is produced (see section 1.2). The identification of the appropriate geological conditions and the necessary surface area is particularly important in the case of an open cycle, though it is necessary for all options. The importance of such conditions is reduced by partial or complete recycling. The reprocessing and reuse of nuclear materials in reactors reduces the volume, long-term radiotoxicity and thermal waste. If the minor actinides are recycled and consumed, the long-term effects on the performance of deep geological storages will be further improved.

In the case of open cycles, as a result of unforeseen delays in the commissioning of the deep storage, the spent fuel may remain in a stockpile at the reactor sites for much longer periods than initially anticipated, even after the shutdown of the reactors and the dismantling of the required fuel handling facilities. This can add costs and require new installations to handle the packaging and ensure the reconditioning of the fuel during long-term maintenance and for packaging in waste packages and depositing once the deep storage is operational.

Monorecycling requires the construction and operation of MOX fuel recycling and manufacturing facilities. Since these costs are incurred at the same time as energy production, large upfront payments are necessary. For countries with limited installed nuclear capacity, the use of contracts for services offered by the owners of existing recycling capacities has been practiced for decades, and today it is offered in France and Russia.

The options for multi-recycling with fast neutron reactors require investments in recycling facilities and higher reactor costs for fast reactors compared to thermal reactors. Both are upfront costs. Significant cost savings for deep storage can be achieved given the considerable reduction it provides in thermal power, volume and the radiotoxicity of waste.

Economies of scale can be an important factor for countries with low installed nuclear capacity and newcomers. The sharing of infrastructure downstream of the cycle, especially for recycling requiring significant investments, is a way to significantly reduce the financial burden.

Regardless of the cycle option chosen, the financial needs must cover all operations for the complete management of spent fuel up to the storage of radioactive waste, and even beyond the closure of the deep storage. The financial needs to be covered in the management of spent fuels must take into account the overall approach, including both the short-term stages as well as the very long-term duration of the entire system or program. These options have different spending profiles and also different risks and uncertainties that will change over time. The open cycle has lower short-term costs, but the uncertainties and risks associated with the changes increase over time. Recycling and cycle closure options have higher short- or medium-term costs. However, their durability characteristics, which include in particular the optimization of waste packages for surface storage, transport, and storage, limit the operations for preparation and deep storage, thus reducing future uncertainties.

The monorecycling of the fuel allows the manufacture of fuels with plutonium or uranium for reprocessing, which allows for savings of about 25% of the natural uranium. In a multi-recycling scenario, in addition to an initial inventory of fissile materials, depleted uranium stocks provide sufficient quantities of fertile materials to power fast reactors almost indefinitely, thus saving 100% of the natural uranium resources.

In general, recycling options offer more opportunities to bring direct and indirect benefits to economic development compared to the open cycle.

Social acceptance is a common question for all the options corresponding to the social acceptance of nuclear energy as a whole. Social acceptance is generally higher when stakeholders are well informed and there is less uncertainty. In this way, the monorecycling option offers additional advantages to generation in the future. Multi-recycling options represent an important potential benefit for future generations by providing a set of technologies that maximize the conservation of natural resources and the reduction of waste. These environmental impacts must be taken into account in the valuation because they represent increasingly important social concerns.

1.1.3. *Economic evaluations*

The management of spent fuel and related waste requires a long-term, multidimensional global systemic approach that takes into account factors such as:

– long-term energy mix plan: installed and future nuclear capacity (including the exit from nuclear power);

– national and international policies and regulations;

– the available resources, including natural uranium, financial resources and technologies/industrial infrastructure;

– social acceptance;

– geology;

– R&D planning of reactors and site remediation;

– valuation/sale of assets after cessation of nuclear production.

The entire nuclear fuel cycle (from the supply of natural uranium, upstream services, fuel supply), and in particular the spent fuel management program (surface storage, transport, reprocessing/recycling of spent fuel in the event that the recycling option is chosen, waste conditioning for placement in deep geological storage), is implemented in stages. All the steps and their interfaces must be considered holistically in order to evaluate the overall technical and economic performances and values. These comparisons are made taking into consideration the specific context of each country and/or each electrical sector.

1.1.3.1. *Theoretical LCOE evaluations of cycle options*

Numerous theoretical studies have been carried out over many years to carry out a general and idealized comparison of options for the cycle of a nuclear power system at equilibrium with a net present value (NPV) calculation (Piet et al. 2007).

As an example, we can cite the NEA study, published in 2013 (OECD/NEA 2013). The results of the study show that the costs calculated for the open cycle option are lower than those of the other options evaluated, though the differences between the three options in the component of the total fuel cycle are within the limits of uncertainty, taking into account the uncertainties around certain input data.

Within the idealized and simplified analyses of NPV type, for which the discounted costs of deep storage appear to be very low considering the implementation time (~100 years after the reactor unloading of the spent fuel), the value of the recycling options is often underestimated, making the option of direct fuel storage slightly more economical (Kim et al. 2015).

These evaluations make it possible to understand the major impacts on the downstream costs of the various options and, more precisely, to identify the cost factors. However, such an assessment cannot simply be transposed into a specific national decision-making context for cycle options. To this end, more detailed and adapted analyses, which take into account the decision-making factors in the valuation, must be developed.

1.1.3.2. *The alternatives in the implementation of the spent fuel management program*

Running contrary to the theoretical evaluations mentioned in section 1.1.3.1, there are a number of decision-making benchmarks in the implementation of the spent fuel management program to be considered. A series of options may be available over time. For example, various spent fuel storage options can be evaluated close to the reactor pool storage period, including examples such as different underwater or dry storage system technologies implemented on reactor site or at a centralized location. These options are evaluated for use as long-term surface storage pending the preparation of waste packages for future storage, or storage pending reprocessing/recycling of spent fuel. The spent fuel management system involves multiple decisions over a long period of time. Many of these decisions include competing objectives or driving criteria, uncertainties and alternative solutions evolving with market conditions and sociopolitical considerations.

The uncertainties and risks to be considered are of different natures. Some are economic, such as the future cost of commercial services or the required investment expenses, and others are technical, such as the management of the extended surface storage period of spent fuels due to the delayed implementation of the installations for packaging spent fuels in waste packages for further storage. The evolution of the market price of natural uranium is also an important factor. Among other things, this will have an effect on the perceived usefulness of fast neutron reactor technology (4th generation) and as such, the timing of its commercial deployment, which is not currently planned until the second half of this century. This factor will be balanced with the benefits on the savings of natural uranium resources and the minimization of long-lasting waste to be managed.

Sociopolitical considerations are significantly important, including the evolution of installed nuclear capabilities, safety and security requirements, the social acceptance of spent fuel management options and radioactive waste management solutions. Last but not least, the financial aspects (capitalization rate, financial rate of return and the financial performance of nuclear reactor operators) will also play an important role in the path that is chosen. The relative importance of these aspects is specific to each context and can change over time, such that the valuation of the different factors differs by country, and the priorities can change over time.

Future developments may also lead to a wide range of alternative solutions using existing partial industrial reprocessing and recycling capacities that may be deployed pending the future development of advanced reactors, such as capacities adapted to reprocess various types of spent fuels in line with fuel design improvements (ATF or others), or the development of advanced fuel concepts making it possible to multi-recycle plutonium in a light water reactor (e.g. Remix or MOX2 fuels) (IAEA 2019).

There are several options, which can be subdivided into different scenarios, that can be implemented during the long management period until an operational deep storage space is ready to accept the spent fuel packaged in waste packages or vitrified canister. The development efforts leading to the provision of improved industrial solutions offer flexibility in the spent fuel management program in order to minimize risks and contain costs, thus improving the overall financial predictability of the complete spent fuel management solution.

Technological developments, changing market conditions and societal trends require a periodic reevaluation of the spent fuel management strategy in order to understand both new opportunities and new challenges.

The developments made in deep geological storage projects have particularly significant financial impacts in the case of the open cycle when the residual operating time of the nuclear reactor is limited.

1.1.3.3. *Advanced evaluation methodologies*

Using a progressive implementation makes it possible to minimize the management costs of spent fuels and to mitigate programmatic risks, prioritizing flexibility to adapt to future development. It also requires valuing long-term goals in short-term decisions.

The future value of an option also depends on the developments made to keep this option open, corresponding to the value to be lost or gained when this option is no longer available. There is an opportunity cost to keeping an option open or closed. Maintaining options has a cost; the decision to limit or abandon an option or an option should be taken into account during the periodic re-evaluation of the strategy, considering the potential difficulties, issues, and risks and opportunities.

From a methodological point of view, we can list the following potential problems:

– an approach recognizing the benefits of limited early disbursements but underestimating future uncertainties or risks;

– an approach to waiting that does not include a regular reassessment of the options that have remained open;

– a simplified economic approach using static models of idealized options.

The issues related to the evaluation can also be explained by:

– avoiding saturation of reactor pools by ensuring a safe and convenient interim storage solution;

– the safe and optimal development of deep geological storage;

– the reduction of environmental impacts and storage allowances;

– ensuring the robustness of the financing selected;

– obtaining the support and trust of the public.

Finally, we can then give the value factors to be taken into account:

– the preservation of natural resources;

– the promotion of a circular economy;

– the contribution to the security of the supply;

– the reduction of the waste to be put in storage to a minimum;

– the optimization of costs for the safe and secure management of spent fuels and waste.

It should also be mentioned that postponing a decision without a regular reassessment can actually lead to the unintentional closing of options. The interest of keeping the option open will depend on the progress of the implementation of the program. Once the final point has been reached, the value of flexibility is likely to be reduced.

The assessment of plausible future scenarios can guide companies or governments and their stakeholders toward appropriate risk mitigation strategies that correspond to the specific context of each nuclear power plant operator, including its cost objectives and exposure to financial risks in dynamic and uncertain market environments (Passerini 2012; Wigeland et al. 2014).

A spent fuel management system that lasts for a very long period of time (beyond 100 years) requires a systemic consideration valuing costs, risks, time, technological developments and social concerns.

In this context, NPV or LCOE type evaluations are no longer sufficient (Havlíček 2008; NNWI 2020) and innovative evaluation methodologies integrating costs, risks, time and options are essential to develop an optimal spent fuel management program for the purpose of:

– adapting to the uncertain socio-political-economic environment of the various stakeholders;

– minimizing the costs and risks of deployment thanks to a progressive implementation, valuing flexibility to adapt to development and future alternatives;

– evaluating long-term goals as part of short-term decisions;

– considering the sharing of infrastructures that would particularly benefit countries with small nuclear programs.

The valuation methods of real options typically make it possible to accommodate these needs in a dynamic analysis specific to the context studied (Jain 2012; Läuferts 2012).

1.1.4. *Deep storage as a preferred solution*

There is no optimal solution that would lead all countries to a common cycle option or to a single spent fuel management strategy. So far, countries have made their decisions based on their history of nuclear energy development and the values they place on the different characteristics of fuel cycle options, as well as their point of view on the long-term use of nuclear energy. Energy policy can also evolve over time, which leads to the transition from a preferred solution to another option. A spent fuel management system extends over a very long period of time (beyond 100 years) and requires a holistic evaluation via the use of innovative evaluation methodologies, beyond simple NPV analyses, which include costs, risks, time and alternatives.

Today, the development of a deep geological storage facility for final waste disposal is considered the best available solution, regardless of the cycle option that is chosen. Access to suitable geological conditions and characteristics and the development time of these installations currently constitute some of the most important issues for all stakeholders at the global level, inducing significant uncertainties in the management programs that must be integrated. Technological developments, including those involving SMRs (small modular reactors) or *advanced modular reactors* and/or advanced fuels, changes in regulations, the evolution of market conditions and societal values require a periodic reassessment of the spent fuel management strategy to understand new opportunities and new challenges. The implementation of management programs is carried out in stages with a holistic assessment of alternatives to multiple decision points via the use of adapted methods taking into account the various dimensions, costs, risks, time and options.

1.2. Costs and methods of financing radioactive waste management

1.2.1. *Introduction*

The management of radioactive waste takes place over the very long-term. It essentially consists of isolating these wastes from the biosphere, containing them and the delay of the dispersion of their radioactive components so as to allow the reduction of their harmfulness by radioactive decay. The duration of radioactive decay is a physical property of matter; it is measured in half-lives, the time necessary for the radioactivity to have decreased by half. A half-life is a characteristic of the physical element; it can be relatively short (8 days for I^{131}, 30 years for Cs^{137}) or long to very long (24,110 years for Pu^{239}, 301,000 years for Cl^{36}).

The methods to be implemented for the management of radioactive waste must therefore allow for the waste to be isolated over very long periods of time. These periods go far beyond what can be controlled on the scale of societies and civilizations. It is therefore necessary to devise a management method that does not call for human intervention over time.

Storage in a deep and stable geological layer has been the subject of a consensus for many decades as to its ability to respond to the challenge posed. After its closing, it can operate autonomously without human intervention for the required periods. Naturally, it is up to the safety studies to confirm this design choice.

Deep geological layer storage in practice forms part of the logic of sustainable development and intergenerational equity as defined at the Earth Summit in Rio in 1992; these two concepts recognize that "development that meets the needs of current generations must be done without compromising the ability of future generations to meet their own", on the one hand, and that it is necessary to "avoid postponing negative consequences (environmental, economic and social) on future generations" of our development model, on the other hand.

In this case, scientific reasoning meshes with the moral duty that current generations have to create a solution for the management of radioactive waste respectful of future generations as soon as possible, and in any case without undue delay.

The option of long-term surface storage clearly does not meet the requirements set out above; even if it can be considered safe on a century-wide scale, there is no guarantee that future societies will have the intellectual and material means to perpetuate it almost indefinitely. Indeed, a surface storage facility differs from a storage facility in that, first of all, the materials or waste is stored there with the intention of recovering them, and second, by contrast, there is no intention of recovering them: the storage is the final stage in the process. The monitoring and maintenance of the stockpile and the materials or waste stored there must therefore be ensured on an ongoing basis and the guaranteed feasibility of transport out of the stockpile.

The integrity of the structures and the operability of the surface storage facilities are inevitably limited in time, though it is projected they will last beyond the end of the century. The surface storage must therefore be renewed periodically and the materials or waste contained therein transferred to a new facility; consequently, surface storage cannot be considered either as permanent or as a definitive solution; a storage system must ultimately be put in place. Ensuring the cost coverage of a procedure that is so uncertain in its scale and carried out over such a long period appears to be almost completely infeasible with the economic tools available: any future updates would annihilate distant costs and inflation would destroy the value of the provisions. Multisecular surface storage thus implies a transfer of resources, both technical and financial, to future generations. This does not respond to the notion of ethical responsibility, that is to say, the obligation we have to be accountable for our actions and to bear their consequences, and therefore will not be discussed here.

The assumption of responsibility by the current generations implies that they assume the costs of the management to be put in place. These costs must therefore be estimated as best as possible and the means to ensure that their coverage is put in place. In this context, section 1.2 is divided into two main sections. The first, section 1.2.2, deals with the analysis of the costs related to the long-term management of radioactive waste. Section 1.2.3 addresses the financing mechanisms that seek to ensure the financial coverage of the long-term management of these wastes.

1.2.2. *Analysis of the costs related to the long-term management of radioactive waste*

1.2.2.1. *Context*

As mentioned in the introduction, in the particular case of the management of long-lived intermediate activity waste (LL-ILW) and HLW, there is a broad international consensus on geological storage in galleries in a host geological formation as an adequate solution for the management of these long-term wastes. This consensus has also manifested itself in the political choice for its implementation in several countries, including Finland, France, Sweden and Switzerland. This solution is also recommended by ONDRAF[1] in Belgium.

In the case of geological storage, most of the developments (research, development and demonstration) concern geological storage in galleries. However, it should also be noted that in parallel with this solution of geological storage in galleries, research has been carried out at an international level on geological storage in deep boreholes for high activity and/or long-life packaged waste (these are essentially conceptual studies). Storage in vertical deep boreholes[2] conceptually consists of the "unloading" of packages of radioactive waste packaged in narrow boreholes with a depth of more than 1,000 m (see Figure 1.6). Most studies call for the storage of waste in boreholes that can reach a depth of 5–6 km. The storage zone, located in the lower part of the boreholes, can reach 2 km and have a diameter of between about 40 and 90 cm. The annular space between the lining of the borehole and the surrounding environment and that between the waste packages and the lining of the borehole can be filled with materials chosen so that the storage system achieves the best possible performance. Horizontal "plugs" are placed at regular intervals in the storage zone in order to reduce both the vertical pressure on the waste packages and the risk of radionuclides and chemical contaminants rising up along the walls of the borehole. The drilling segment located above the storage area must be adequately closed. After the borehole has been completely closed, the system, in theory, will safely and passively confine and isolate the radioactive waste for as long as is necessary.

The main advantages of deep drilling are as follows:

– the lack or near lack of groundwater movement at the depths at which the waste is stored, and the low probability of contact with the groundwater system closer to the surface, which itself is connected to the biosphere;

1 National organization for the management of radioactive waste and fissile materials (Belgium).

2 Research is also being carried out on the technology of so-called "deviated" deep drilling, in which the lower part of the drilling is (sub-)horizontal. This deviated deep drilling is commonly used in the oil industry.

– the significant thickness of all the geological layers that isolate the waste from the biosphere;

– the impossibility, or at a minimum the extreme difficulty, of recovering the waste after the complete closure of the boreholes (for the types of waste that it would be desirable to never be able to recover, such as for reasons of non-proliferation).

The main disadvantages of deep drilling are the following:

– according to some opinions, the storage system is essentially a mono-barrier system, where the only real barriers are geological (the host formation and some of the upper geological layers);

– the relatively limited diameter of the packages that can be put in storage;

– the impossibility, or at least the extreme difficulty, of recovering the waste after the complete closure of the boreholes.

Although this area continues to evolve, the concepts of geological storage in underground boreholes remain significantly less mature than the concepts of geological storage in galleries, both from the point of view of the technologies to be implemented and that of the scientific capacity to demonstrate that such storage systems are able to protect humans and the environment from the risks posed by waste for as long as necessary.

Box 1.1. *Advantages and disadvantages of deep drilling*

According to the polluter-pays principle, producers of radioactive waste must generate sufficient income starting from the present time to cover future needs in terms of financing geological storage facilities. The period of time between the generation of income by waste producers and the first expenses generated by the development of geological storage can range from a few decades to more than a century. This significant duration requires establishing a cost estimate that is reliable in order not to transmit the financial burden of the long-term management of radioactive waste to subsequent generations. This results in the need to develop a cost calculation methodology that makes it possible to obtain a transparent, traceable and robust cost calculation and that can take into account all the uncertainties related to this calculation and the geological storage program. These uncertainties stem from various causes, in particular technological issues or those related to the degree of maturity of the program[3] for geological storage.

3 Program: The term "program" refers to the coordination and execution of all the projects that are necessary for the construction, operation and closure of the geological storage facility.

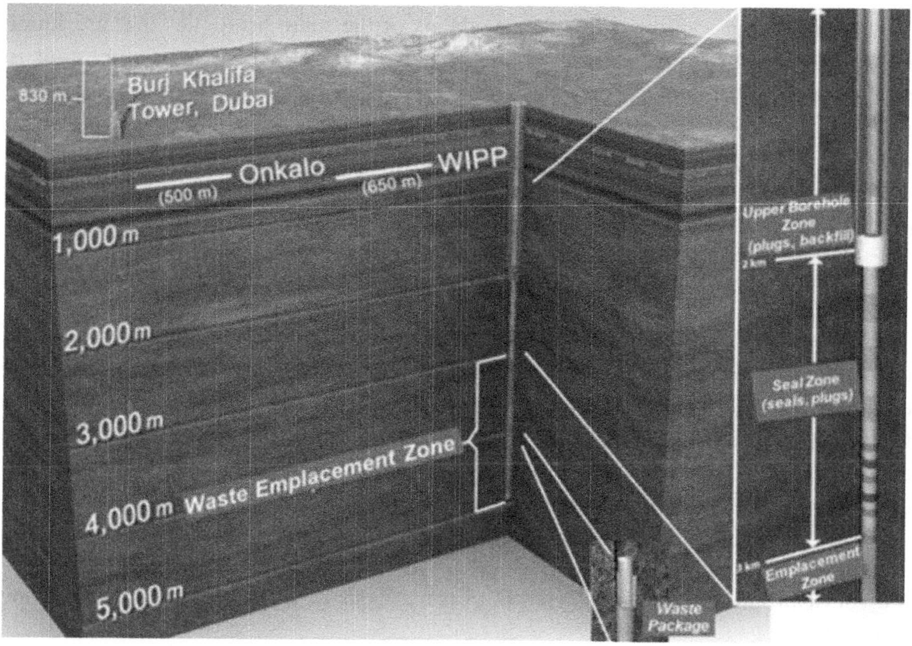

Figure 1.6. *Principle of geological storage in deep drilling (source: Freeze et al. 2019). For a color version of this figure, see www.iste.co.uk/percebois/nuclear2.zip*

In this context, many countries have developed their own methodology by integrating input from past experiences (feedback) from projects related to the nuclear industry (construction or decommissioning of new power plants) and megaprojects[4]. These methodologies, although each presenting different specificities, share a similar approach and are built according to recurring patterns.

The first section of this chapter presents the main principles that can be found in the cost analysis methodologies applied in different countries (Belgium, Finland, France and Switzerland in particular) as well as in numerous international guidelines (such as NASA 2015; OECD/NEA 2017; GAO 2020; IAEA 2020).

4 "Megaprojects" refer to projects with a total cost of more than 1 billion euros. These megaprojects are found in different sectors: infrastructure, nuclear, civil aviation, etc. A detailed analysis of the REX of megaprojects, conducted among others by Flyvbjerg (2014), showed that an overspending of 50% with respect to the initial budget was frequent and that overspending of more than 50% was not uncommon.

The following sections present a synthesis of the main principles that are found in the cost analysis methodologies referenced above. However, it is important to note that there are many methodologies that have been developed in different countries and for different types of projects and which, even if they are based on similar principles, may differ in how they are applied specifically. The following sections also present certain specific considerations implemented in the methodologies developed by ANDRA[5] (France) and the ONDRAF (Belgium).

1.2.2.2. *Generic methodology for analyzing costs*

The purpose of this cost analysis methodology is to determine the costs of implementing geological storage, without taking into account inflation and price adjustments. Inflation and the adjustment rate are discussed in the next section on financing mechanisms. The costs, excluding inflation and adjustments, are the non-adjusted gross costs (also referred to in some countries as *overnight* costs, in other words, as if the project were carried out instantly).

The analysis of the costs for geological storage, and in general for any industrial project or program, must be carried out on the basis of a transparent, trackable and robust process:

– Transparency: each stakeholder (waste producers, the organization in charge of waste management, decision-makers and the public) can share the same understanding of the project and its cost analysis, which allows for increases in the mutual trust of the various actors.

– Trackability: each hypothesis is identified, and each data used are justified and their origin is clearly documented. This traceability is a prerequisite to guarantee the quality of the estimate (avoiding double counting and omissions among other things) and its credibility.

– Robust: the cost estimate obtained by applying the cost analysis methodology can cover the impact of minor or even more substantial modifications to the program.

The cost analysis methodology is an iterative process, as illustrated in Figure 1.7. This process can be divided into 12 steps, which are divided into four phases.

5 National agency for the management of radioactive waste (France).

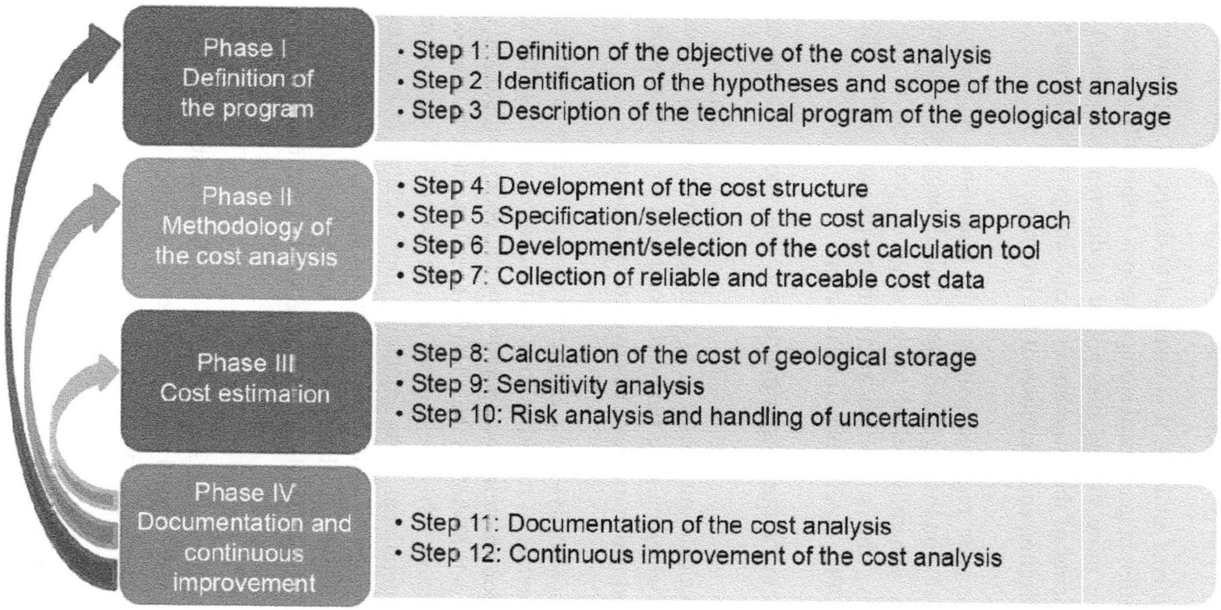

Figure 1.7. *Iterative and sequential process for cost analysis. For a color version of this figure, see www.iste.co.uk/percebois/nuclear2.zip*

1.2.2.3. *Phase I – definition of the program*

1.2.2.3.1. Definition of the objective of the cost analysis (step 1)

Before starting the cost analysis, it is essential to define its objective. This may be a comparison of different geological storage options (especially options related to the selection of one or more storage sites and/or in different host rocks), a budget estimate for the financing of the program, or a cost analysis that forms part of the monitoring of the realization and execution of the program.

The degree of precision required will then be established on the basis of the objective sought, but also on the degree of the technical maturity of the project. This degree of maturity is generally linked to the stage of progress of the program: conceptual, feasibility, the simplified preliminary project (SPP), the detailed preliminary project (DPP), procurement, or execution.

1.2.2.3.2. Identification of the assumptions and the scope of the calculation (step 2)

The scope of the calculation allows us to distinguish between the costs included in the calculation and which must be financed, along with the costs outside the scope of such calculation and which must not be covered by financing.

The scope of the cost analysis of a geological storage facility covers not only the investment costs as well as all the other costs that are necessary from the time of the authorization of the construction and operation of the geological storage (T_0)[6] until its closure (T_C) (the costs related to the post-closure phase [institutional control] are also included) (see Figure 1.8)[7]. These costs include the cost of engineering studies, construction costs, operational costs (including those related to maintenance and renovation)[8] and the costs of closing the storage facility (including operations to dismantle surface facilities). In other words, the cost analysis of a geological storage requires a complete analysis of the costs of the lifecycle of the program (lifecycle cost analysis).

The scope of the calculation, supplemented by all the assumptions necessary to calculate the costs of a geological storage project, forms the reference scenario. The reference scenario is critically important in the calculation.

---

6 In general, the costs prior to the authorization for the construction and operation of the geological storage are not included within this scope. These costs are financed by other mechanisms.

7 There may be a delay between the end of the storage and the closure of the geological storage (oversight and monitoring phase).

8 Rejuvenation refers to the renewal of equipment during the period of operation.

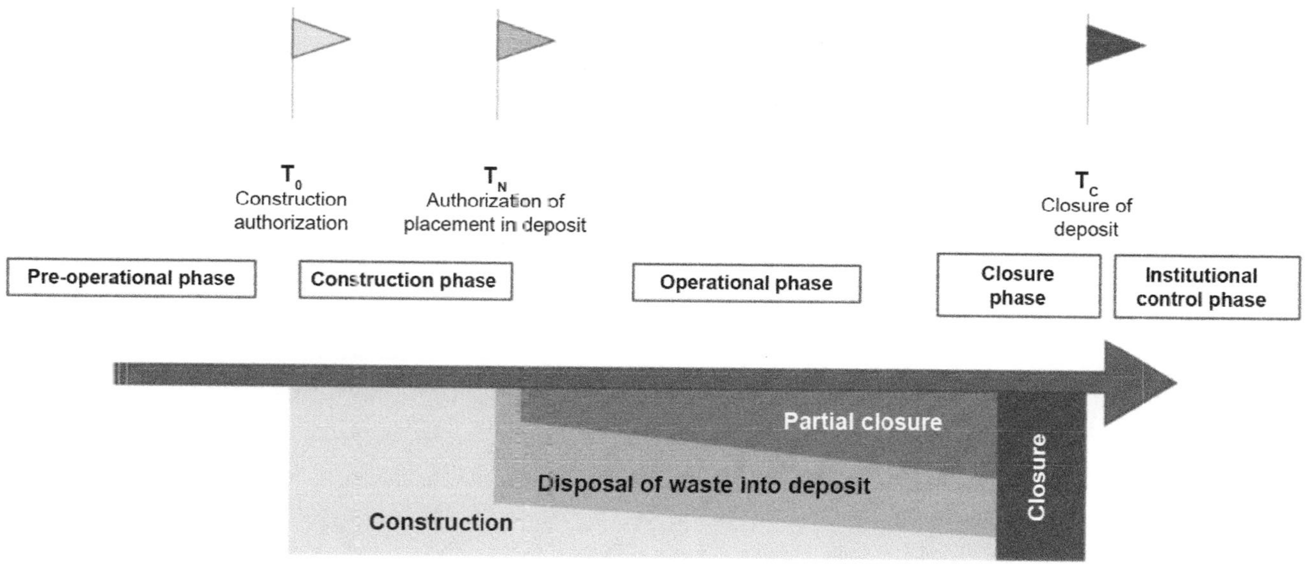

Figure 1.8. *The temporal phasing of a geological storage and the schematic evolution of activities during the program: construction, disposal of waste, partial closure of the storage galleries and final closure. For a color version of this figure, see www.iste.co.uk/percebois/nuclear2.zip*

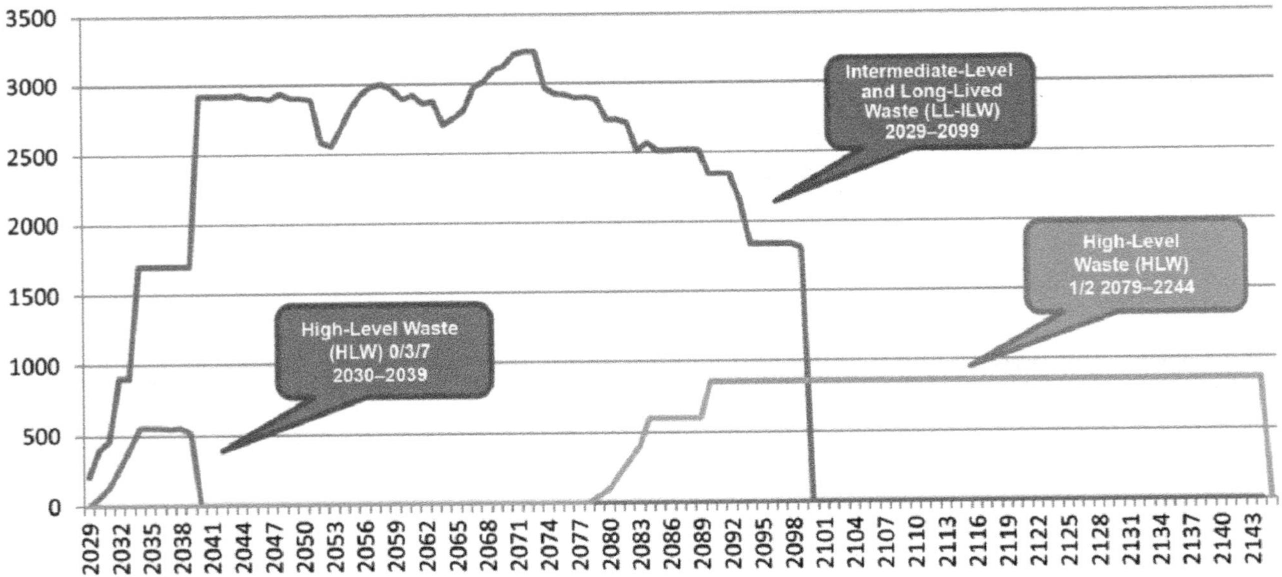

Figure 1.9. *Chronic delivery of radioactive waste (intermediate-level and long-lived waste [LL-ILW] and high-level waste [HLW, Cigeo geological storage program]; source: ANDRA 2014). For a color version of this figure, see www.iste.co.uk/percebois/nuclear2.zip*

The reference scenario brings together:

– the legal framework;

– the regulatory requirements, established by the nuclear safety authority and other competent authorities (e.g. mining regulations);

– the assumptions about the location of the storage (for the least advanced programs);

– the hypotheses related to the construction, the operability and the closure operations of the geological storage, as well as the determination of the phases of these different tasks;

– the inventory of the primary packages of radioactive waste, as well as their delivery rate on the site of the storage facility. An example of a forecast delivery schedule for radioactive waste for the French geological storage program under the responsibility of ANDRA (Cigéo project) is provided in Figure 1.9.

1.2.2.3.3. Description of the geological storage technical program (step 3)

This technical description describes the geological storage facility, as well as the main phases from the start of construction to the closure of the storage facilities (geological storage planning). The technical description consists of calculation notes, studies, and plans carried out by the engineering offices. The degree of maturity of the technical file depends on the progress of the development of the technical program.

1.2.2.4. *Phase II – cost analysis methodology*

1.2.2.4.1. Development of the cost structure (step 4)

The analysis of the costs requires a division of the program into sub-programs, which are themselves divided into projects and tasks. The costs associated with these tasks are the cost elements of geological storage. This approach makes it possible to structure the cost of the entire program into cost elements that together cover all activities related to the storage. This division into cost elements provides a cost structure referred to as the work breakdown structure. The breakdown structure of the project has different levels, the highest level (level 1) representing the entire geological storage program. In general, the cost reduction is carried out, on the one hand, on the basis of the temporal phasing which is common to all geological storage programs and, on the other hand, on the basis of the major substructures of the storage facility (surface installations including radioactive waste conditioning installations, surface underground connections and underground installations). An example of a work breakdown structure is provided in Figure 1.10.

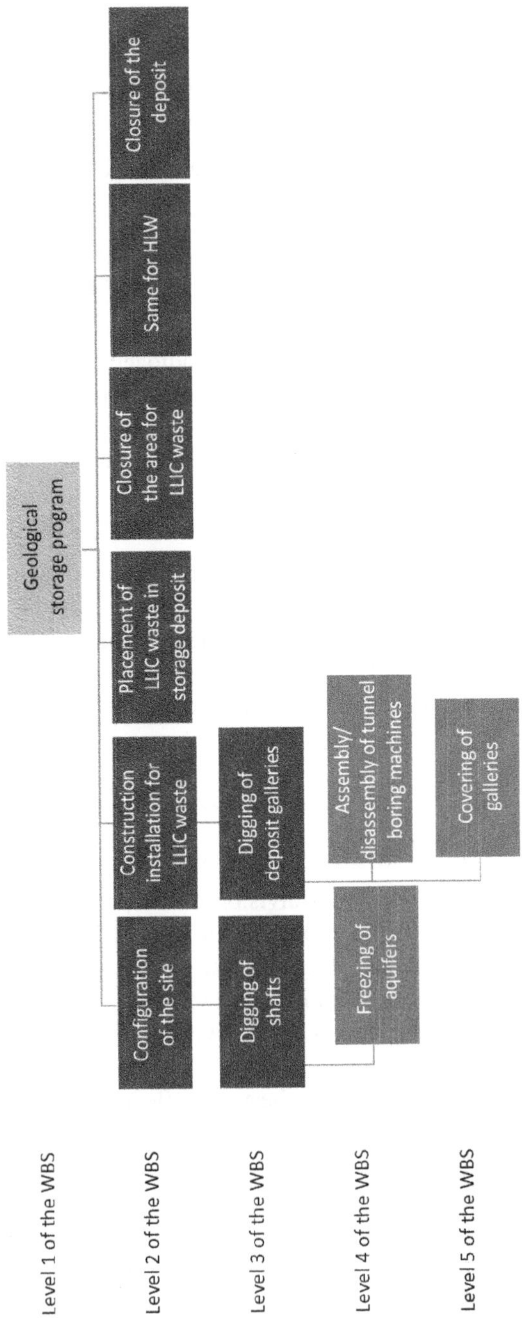

Figure 1.10. *Work breakdown structure (WBS) of the storage program. For a color version of this figure, see www.iste.co.uk/percebois/nuclear2.zip*

1.2.2.4.2. Specification/selection of the cost analysis approach (step 5)

The cost estimation of the geological storage program is carried out at the "lowest level" of the project management structure (the cost elements). These costs are generally obtained according to different approaches which vary according to the degree of maturity of the cost element. The most commonly used approaches are the following:

– Detailed approach: the costs of each element of the breakdown structure of the project are determined on the basis of plans, technical specifications, quantities and unit prices.

– Analogy approach: the cost of a system, a structure or a component is taken over from a similar element whose costs have already been determined in another project.

– Parametric approach: the cost is determined based on a historical database of similar systems from which a statistical analysis can be carried out in order to deduce a correlation between the costs and the main parameters of the system.

Regardless of the approach taken, it is important for the cost estimate to be transparent and traceable in order to avoid omissions, duplicates or the introduction of hidden provisions in the costs. Special attention must be paid to these requirements, in particular through the implementation of quality audits and peer reviews.

1.2.2.4.3. Development/selection of the cost calculation tool (step 6)

When the breakdown structure of the project has been set and the cost analysis approaches have been chosen, the estimation tool can either be developed or selected from software specifically designed for cost analysis. This is an important step that must be carried out according to the needs, but also according to the available resources.

1.2.2.4.4. Collection of reliable and traceable cost data (step 7)

The cost estimation software must now be supplied with reliable data. During this step, the important elements are:

– the collection of data from historical databases;

– the data must come primarily from primary sources of information and not be calculated/derived from other sources;

– the data must be clearly understood and documented, particularly so as to ensure their traceability;

– consideration should be given to the serial effect which leads to a reduction in costs for repetitive standard operations;

– the origin of the sources and all the calculations necessary to determine the costs (in the case of a secondary source) must be documented.

As mentioned previously in several instances, it is essential to document this step as completely as possible in order to ensure the traceability of the data.

1.2.2.5. *Phase III – cost estimation*

1.2.2.5.1. Calculation of the cost of geological storage (step 8)

From the work breakdown structure and the cost data, the work schedule cost can be calculated by totaling the lowest level elements of the work breakdown structure.

1.2.2.5.2. Sensitivity analysis (step 9)

The determination of the cost of geological storage established in the previous step cannot be used on its own, because it does not make it possible to identify the parameters that have the greatest impact on the final cost, nor the way in which these parameters influence the final cost. For this reason, sensitivity studies must be carried out.

Sensitivity studies make it possible to do the following:

– show the various stakeholders that the calculation can withstand changes in assumptions;

– identify the assumptions and/or cost elements that can have the most impact and which therefore require future developments (R&D, engineering, prototypes, etc.).

The most important parameters for sensitivity studies are identified by considering those that:

– have a significant impact on the program schedule;

– form a significant relative percentage in the estimated cost;

– involve a decrease/increase in the volume of services or a variation in hourly costs;

– lead to a change in the assumptions or boundary conditions.

Some examples of the parameters considered in sensitivity studies are:

– the change in costs of construction materials (steel, concrete);

– changes to the design;

– recoverability requirements (the recoverability of the storage of radioactive waste, that is, the possibility of recovering radioactive waste after its storage is required in many countries).

The sensitivity study can also be carried out through the analysis of risks and opportunities of a project (see step 10 described below).

1.2.2.5.3. Risk analysis and handling of uncertainties (step 10)

Accounting for uncertainties is a critical step in establishing the costs of a program, and more particularly in the context of a geological storage program for which there are many uncertainties, particularly given the time scale of the program. These uncertainties can have several origins:

– uncertainties related to technological maturity (the uncertainties will be greater for a newer technology (prototype) than for standard technologies in the industry);

– uncertainties related to the maturity of the project (the uncertainties are greater at the beginning of the development phase of a program while they are reduced with the advance of studies);

– risks and opportunities that may impact the planning and/or investment costs.

Taking the uncertainties into account requires an appropriate strategy to address them and an analysis of their risks and opportunities.

As mentioned in the previous step, the sensitivity study can be incorporated into how uncertainties are addressed and the analysis of risks/opportunities. While the sensitivity study makes it possible for a single parameter to be varied at a given time, the analysis of uncertainties and risks/opportunities makes it possible to consider the impact of all the uncertainties of a program in a single calculation.

The treatment of uncertainties and the analysis of risks/opportunities generate the considerations of additional costs for the risks and a reduction in costs for the entities.

In order to quantify these, it is important to define a transparent and unambiguous terminology of the various elements that constitute the final cost of the program. This final cost consists of the basic costs and the additional costs which cover the operating uncertainties and the risks. This is illustrated in Figure 1.11, for which the elements are described below.

1.2.2.5.4. Base cost

The base costs cannot include financial provisions to cover possible consequences of a risk, unless these are clearly identified as such in the base cost (risk mitigation), or are provisions to cover uncertainties at the level of the definition and analysis of the cost elements of the program. In other words, the base cost is a cost without any provisions for "hidden" costs, in order to guarantee the transparency of the calculation. This cost is sometimes referred to as an optimized cost, insofar as this cost corresponds to that of the program that would take place in accordance with the assumptions and the initial schedule.

As noted earlier, the base cost may contain provisions for risks, provided that these are clearly identified, and only for risks whose consequences are considered unacceptable (in terms of cost and/or schedule). For the latter, it is then preferable to provide additional content to the basic program in order to reduce these risks. This additional content corresponds to the mitigation of the risk. The reduced risk can then be included as a residual risk in the subsequent risk analysis (see below for the risk analysis).

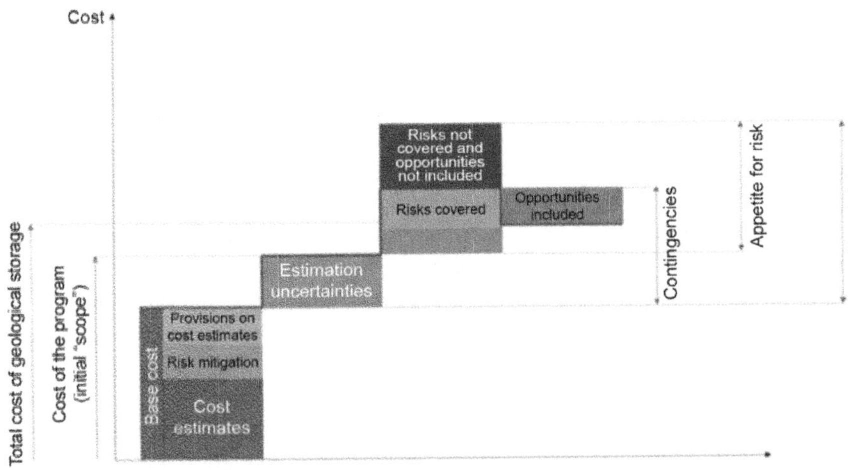

Figure 1.11. *Structuring of the cost elements constituting the total cost the geological storage program. For a color version of this figure, see www.iste.co.uk/percebois/nuclear2.zip*

The provisions for the estimated costs (or allowances) are included in the base cost and cover expenses that cannot be defined at the time of estimation and are based on the observations made by the cost estimators. These provisions cover the project activities (technical, engineering and execution) (e.g. costs related to losses during the breakdown of materials).

As mentioned above, the calculation of the basic costs is carried out at the lowest level of the breakdown structure of the project.

1.2.2.5.5. Contingencies

Contingencies are costs incurred in addition to the base costs, and which cover:

– The uncertainties related to the estimation of the basic costs. These are the uncertainties of estimation.

– Events characterized by a certain likelihood of occurring and by a negative or positive impact on the cost, and which are taken into account in the cost analysis of geological storage costs. These are the risks and the opportunities, respectively.

Contingencies are associated with a given program scope. The contingencies cannot in any way be considered as a provision, making it possible to cover variations in the scope of the geological storage program.

The estimation uncertainties, as well as the risks and opportunities are detailed in the following two sections.

1.2.2.5.6. Uncertainty of the estimation

The uncertainties in the estimation are uncertainties included in the basic scope of the program. They involve events related to the execution of the program (e.g. equipment failure, weather hazards or logistical delays), which lead to an increase in costs and/or an extension of the schedule.

The level of the estimation uncertainties depends on the level of maturity of the program and the variability of the cost estimates (standard equipment with low variability up to the prototype for which the feedback is low, or even zero).

It is important to note that the estimation uncertainties do not contain provisions for a potential change in the reference scenario (outside the scope of the program), such as the impact of a regulatory change, a major modification of the contract, a catastrophic event (*force majeure*) or financing difficulties. These changes must be studied as part of an analysis of the risks that are outside the scope of the reference scenario.

The estimation uncertainties can be calculated at different levels of the work breakdown structure:

– At the highest level of the work breakdown structure, that is, at the program level. This is the simplest approach. This approach can be used, for example, when a fixed percentage of the total cost of the project is added in order to provide for a complementary financial provision.

– At the level of the group of activities (civil engineering, underground construction, ventilation, mechanical equipment, etc.): this approach was taken by ANDRA in the Cigéo project.

– Application at the lowest level of the work breakdown structure. This approach has been taken in particular by ONDRAF for the Belgian geological storage program.

The estimation uncertainties can be obtained according to two possible approaches: the deterministic approach and the probabilistic approach.

A deterministic approach is one in which a fixed percentage is added to the base cost. The value of this percentage is usually deducted from the experiences shared. For instance, an example of the application of this type of methodology is presented below (Parsons 1999). In this example, the fixed percentage results from two components, one linked to the degree of maturity of the project (conceptual stage, feasibility, DPP or execution) and the other linked to the degree of technological maturity (ranging from new technology to standard technology in the industry). In this approach, the following percentages are considered for project uncertainties and technological uncertainties, respectively (see Table 1.1).

Maturity of the project	Percentage	Technological maturity	Percentage
Concept	30–50	New concept	>40
Feasibility	15–30	New concept with pre-study	30–70
Detailed preliminary project	10–20	New concept with advanced study	20–35
Execution	5–10	Concept derived from an existing concept	5–20
		Standard concept in the industry	10

Table 1.1. *Project maturity and technological maturity*

Although these deterministic approaches are still used today, insofar as they benefit from decades of experiences and have the advantage of being relatively simple to apply, probabilistic approaches are being implemented with increasing frequency, as is the case at ANDRA, NAGRA[9], the ONDRAF, or SKB[10]. A lot of research has been carried out in the field of probabilistic approaches in recent years, which has now lead to a high proficiency in cost analyses using these methods (such as the Monte Carlo-type probabilistic analysis). The advantage of such methods is to introduce ranges of uncertainties at the level of the groups of activities (see above) or at the level of the lowest elements of the work breakdown structure. These ranges of uncertainties are given specific values through probability distributions for the cost elements of the work breakdown structure. The probability distributions that are generally chosen are either uniform or triangular probability distributions. It should be noted that some more specific distributions (like the Beta distribution) are sometimes used.

Triangular distributions are used for cost elements of the work breakdown structure, for which a certain degree of confidence is placed in the cost estimates. Otherwise, a uniform distribution is used.

The triangular probability distributions (Figure 1.12) are fully defined by the specification of three characteristic values:

– the minimum value of the range of variation of the cost element, also called the optimistic value;

– the most likely value of the cost element, sometimes given as the *Best Estimate*;

– the maximum value, also known as the pessimistic value.

In general, the triangular probability distribution is asymmetrical. In practice, real-world experiences show that assessors and program managers have a tendency to underestimate costs (known as the optimism bias)[11]. As a result, the probability that the cost will be higher than the most likely estimate (running over budget) is greater than the probability that the cost will be lower than the most likely value (an underestimation). These three characteristic values are set by the cost estimators and are a function of the degree of maturity of the project, as well as the level of confidence of the estimators.

9 Swiss agency of radioactive waste management.

10 Swedish agency of radioactive waste management.

11 The optimism bias has been the subject of numerous studies (e.g. see, Flyvbjerg and COWI 2004).

The uniform probability distributions (Figure 1.13) are fully defined by the specification of two characteristic values: the minimum value of the range of variation of the cost element (optimistic cost) and the maximum value (pessimistic cost).

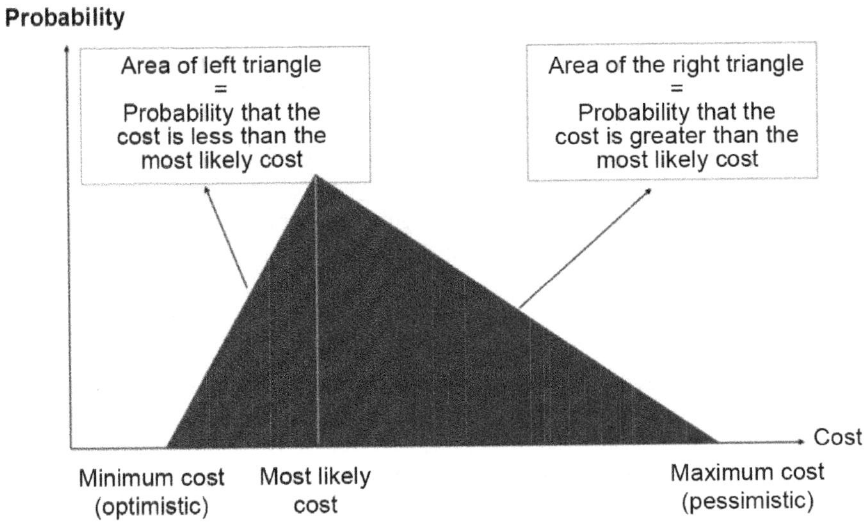

Figure 1.12. *Triangular probability distributions. For a color version of this figure, see www.iste.co.uk/percebois/nuclear2.zip*

Figure 1.13. *Uniform probability distribution. For a color version of this figure, see www.iste.co.uk/percebois/nuclear2.zip*

The characteristic values of the probability distributions are set on the basis of the feedback from other projects and/or on the basis of expert judgment.

In France, ANDRA follows this approach in calculating the estimation uncertainties for the estimation of costs. The probability distributions and their characteristic values established by ANDRA have been established for the total cost by "professions" (activity groups). Some typical examples taken from ANDRA calculation in 2014 are given in Table 1.2 (ANDRA 2014).

Professions	Shapes	Characteristic values
Mechanical equipment – nuclear installation	Triangular	–10 to +20% of the estimated cost
Mechanical equipment – nuclear *process*	Triangular	–20 to +20% of the estimated cost
Underground technical equipment	Triangular	–30 to +50% of the estimated cost
Underground works – tunnel boring machine	Uniform	–10 to +10% of the estimated cost
Human resources – operations	Uniform	–10 to +10% of the estimated cost personnel

Table 1.2. *Probability distributions*

In Belgium, the ONDRAF also selects uniform and triangular distributions for the cost elements of the breakdown structure for the project. On the other hand, some elements of the work breakdown structure are not affected by a probability distribution, because they are a function of other cost elements (e.g. engineering costs are estimated on the basis of a percentage of investment costs). This functional dependence leads to an implicit correlation between these elements of the breakdown structure of the project. The characteristic values of the probability distributions are either extracted from the cost databases and supplemented by the judgment of experts, or obtained by applying a procedure developed by the ONDRAF and which is based in part on the documents of the Association for Advancement of Cost Engineering International (AACE 2019). This procedure allows for the characteristic values of the probability distributions to be set on the basis of the following two parameters:

– The degree of maturity of the cost element: This degree of maturity is quantified in the form of the classes defined by the AACEI. These classes range from 5 to 1, with class 5 representing the conceptual progress stage of the project and class 1 referring to the procurement stage before the project's execution. The choice of classes is first proposed by the estimators and is then analyzed during workshops bringing together multidisciplinary experts appointed by ONDRAF and waste producers. For each of these five classes, the AACEI determines a high range and a low range, in which the real cost of the project (contingencies included) is to be located (AACE 2008). These ranges are expressed in terms of the percentage of the total estimated cost (the sum of the base cost and contingencies) and are produced from the analysis by the AACEI of the REX of a very large number of projects. The upper range represents the probable costs of the program that exceed the estimate (a budget overrun). By contrast, the low range represents an overestimation of costs. These low and high ranges are shown in Figure 1.14 (in the blue area). The AACEI considers that 80% of the projects carried out fall within these ranges.

– The confidence level: This is the responsibility of the cost estimator and is quantified on the basis of a judgment. This confidence level can be low, medium or high, and it is given the following color code: red, orange and green.

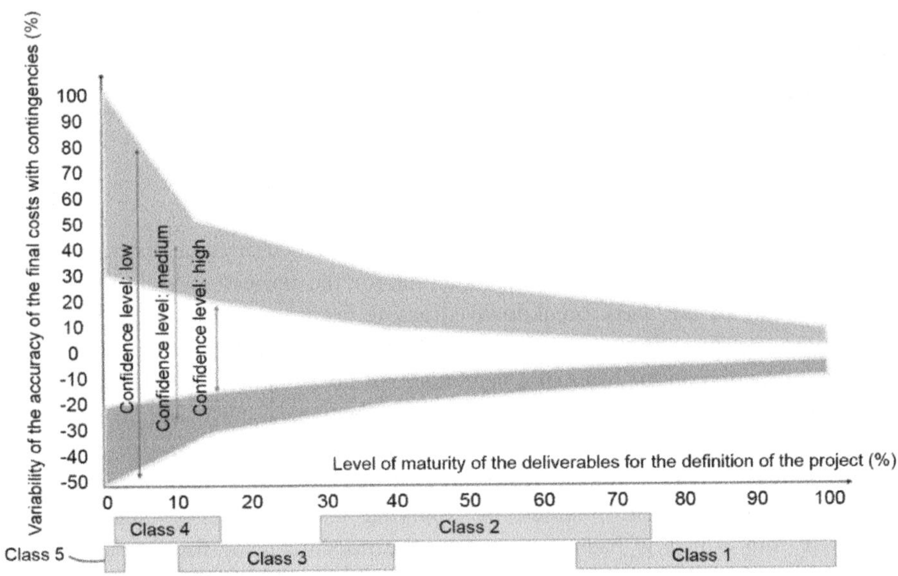

Figure 1.14. *Variability in the accuracy of estimated costs by AACEI class. For a color version of this figure, see www.iste.co.uk/percebois/nuclear2.zip*

The degree of maturity of the cost element (expressed by the AACEI class) and the confidence level are then combined to determine the ranges of uncertainties of each cost element. This is illustrated in Figure 1.14. In the approach taken by the ONDRAF, it is also assumed that the base costs represent the minimum values (optimistic values). By definition, the basic costs do not present contingencies.

This procedure has the advantage of being structured and easily applicable by cost estimators. It also facilitates exchanges between experts during the *workshops*, the AACEI classes that allow the arguments for each one to be justified. At the end of this procedure, the characteristic values of the probability distributions are obtained. The cost estimators check and validates these ranges with regard to their experience and databases.

When the characteristic values of each probability distribution are challenged, it is necessary to take into account the interactions between the different elements of the costs. There are certain cost elements are correlated with other ones. For example, if the cost of concrete is a critical parameter for several cost elements, it is therefore necessary to take into account this correlation between the cost of concrete and these different elements. This is taken into account by defining the correlations between the different probability distributions. Correlations are generally defined in the form of Pearson correlations.

It is important to note that some software dedicated to simulation calculations by the Monte Carlo method implement other types of correlations (such as the Spearman correlation). The Pearson and Spearman correlations do not reflect the same concepts[12]. It is therefore essential to verify the correlation type when defining these correlations, and also to verify that the selected type represents the needs of the calculation.

Taking into account the correlations is important for accurately capturing the probability distribution curve of the estimated total cost.

Then, when all the probability distributions of the various cost elements (the shape of the distributions and the characteristic values) and the correlations between these elements have been defined, a Monte Carlo calculation is made to establish the curve of the cumulative probability distribution of the cost including the estimation uncertainties (the sum of the base cost and the estimation uncertainties).

12 The Spearman correlation determines a correlation that is not between the values taken by the two correlated cost variables, but between the ranges of the values taken by these two variables.

A schematic form of the Monte Carlo calculation is shown in Figure 1.15. This type of calculation is carried out by ANDRA, the NAGRA and the ONDRAF, among others, though with certain differences between them, particularly in terms of processing correlations.

The cumulative probability curve obtained is shaped in the form of an "S". This S-curve, for each cost on the y-axis, gives the probability that this cost will not be exceeded (on the x-axis).

The ONDRAF uses this S-curve to determine the cost of geological storage: on the basis of a probability (percentile) set by the ONDRAF, the S-curve is used to determine the cost of geological storage to be provisioned in such a way that the final cost of storage is lower than this provisioned cost, with a probability equal to the selected percentile.

A discussion on how this percentile is selected will be presented later in this chapter.

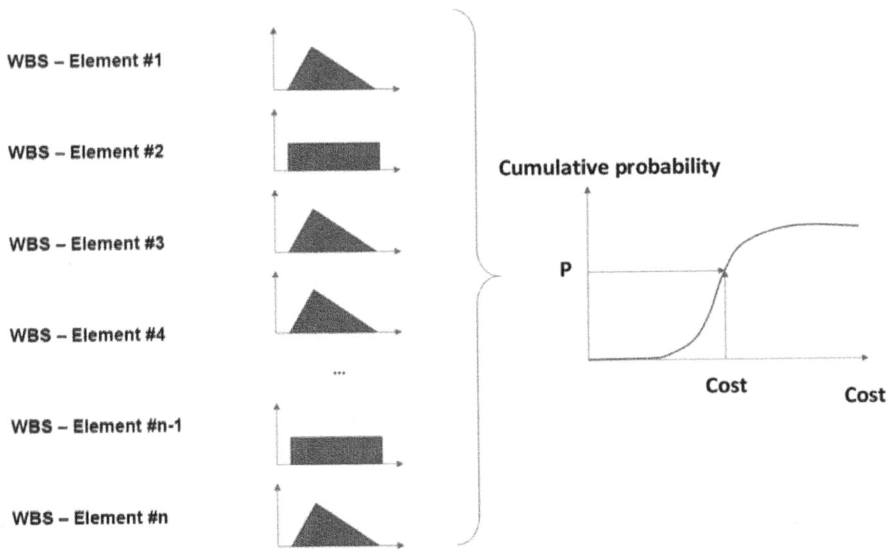

Figure 1.15. *Cumulative probability distribution (basic cost + estimation uncertainties). For a color version of this figure, see www.iste.co.uk/percebois/nuclear2.zip*

1.2.2.5.7. Risks and opportunities

Risks and opportunities are events characterized by a probability of occurrence and by their negative (or positive) consequences to the cost of the program. Risks and opportunities may or may not be the subjects of financial provisions. The risks and opportunities included in the cost calculations are part of the reference scenario.

The risks and opportunities that are the subject of financial coverage generate additional costs that are added to the cost of the program, which consists of the basic cost and the estimation uncertainties.

The risks and opportunities are identified according to several complementary approaches: the analysis of technical documents, and the observations from experience made during workshops that bring together experts from different disciplines.

The risks and opportunities are listed in a register of risks. This register gives the probability of occurrence and the impact of each risk and each opportunity, expressed in terms of cost and/or time delay/savings.

It is also important to check that the risks and opportunities are not already covered by the estimation uncertainties. In this case, they must be clearly identified as such in the register and not be taken into account in the financing provisions in order to avoid being counted twice.

The financial provision (not discounted) to cover the risks and opportunities identified in the risk register can also be calculated on the basis of a probabilistic approach (a Monte Carlo calculation). This approach has been adopted by ANDRA, ONDRAF and NAGRA, among others. The probabilities of occurrence and the impacts of risks/opportunities are modeled by probability distributions. The impacts are generally modeled by uniform or triangular distributions.

Particular attention must also be paid to the definition of correlations between risks/opportunities. Certain specific risks/opportunities are negatively correlated "anti-correlated". In some cases, if one risk materializes, then another risk from the register cannot occur at the same time. It should be noted that there are differences in the approach of ANDRA and that of the ONDRAF regarding the treatment of risks and opportunities. This is in particular due to the degree of progress of the Cigéo project, which is higher than that of the Belgian program.

The probabilistic calculation also gives an S-curve for the cumulative probability curve of risks and opportunities.

In Belgium, as is the case for the calculation of estimation uncertainties, the level of risks and opportunities for which a financial provision is to be expected is fixed by the selection of a percentile by ONDRAF.

The undiscounted provision for risks and opportunities is then added to the sum of the base cost and the estimation uncertainties calculated in the previous step in order to obtain the total amount to be provided for the geological storage of radioactive waste.

Regarding the choice of the percentile, and for informational purposes, the AACEI (Association for Advancement of Cost Engineering International) recommends a percentile of P50. This value is considered as a balance value in the context of managing a sufficiently large portfolio of projects. Some projects will offer the choice of a P50 together with an underestimation of the costs, while others will be overestimated. In this context, the portfolio will be in balance as a whole. In riskier projects, GAO[13] recommends selecting higher percentiles (P70 or P80).

The ONDRAF uses the P50 percentile, while the NAGRA uses higher percentiles. This choice is guided by the willingness of the organization to take risks, but also by the degree of maturity of the geological storage program.

1.2.2.6. *Phase IV – documentation and continuous improvement*

1.2.2.6.1. Documentation of the cost analysis (step 11)

The cost analysis file must include:

– the reference scenario;

– the inventory of the radioactive waste and the time scale for this waste;

– the technical description of the geological storage;

– the cost data and their references;

– the methodology and the result of its application (sensitivity studies and S-curves);

– the result of the risk and opportunity analysis, as well as the risk register.

1.2.2.6.2. Continuous improvement of the cost analysis (step 12)

As in any project and program, the cost analysis is an iterative process. This is especially true for geological storage activities, for which the operational lifespan

13 GAO: US Government Accountability Office.

can last up to a century and more. It is also of primary importance to keep the risk register regularly updated.

1.2.2.7. *International comparison of geological storage costs*

The objective of this section is not to compare the absolute costs of the different geological storage options. Indeed, there are significant differences between the inventories of radioactive waste, the host rocks conserved for storage and, in general, the scope of the programs used by different countries. However, it is interesting to compare the relative shares of each cost element with respect to the base cost (uncertainties of use and risks and opportunities). This comparison between several countries in Europe (Belgium, France, Sweden and Switzerland) is provided in Figure 1.16.

This figure shows that the cost division is similar for the different geological storage projects considered in this comparison. Moreover, this comparative analysis also shows that the probabilistic estimation of contingencies is implicated in many methodologies.

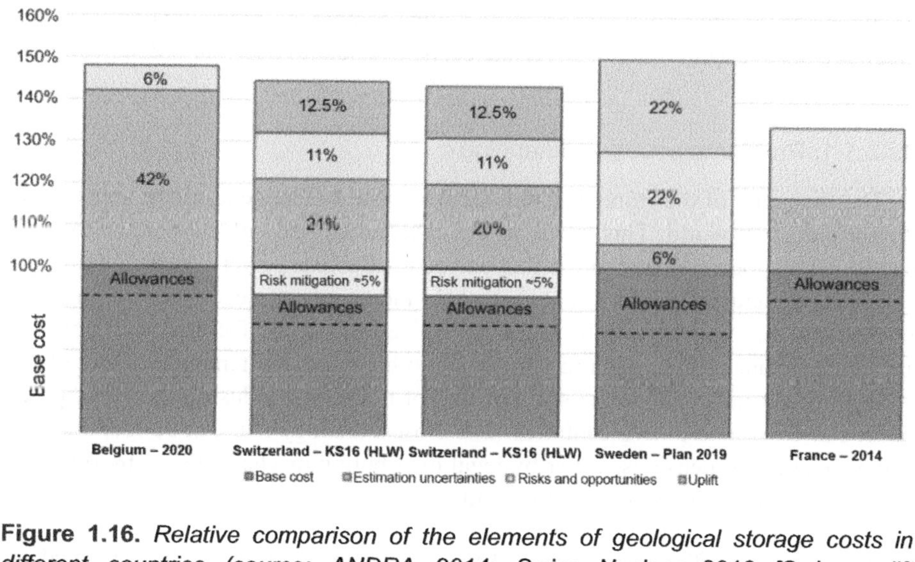

Figure 1.16. *Relative comparison of the elements of geological storage costs in different countries (source: ANDRA 2014; Swiss Nuclear 2016 [Swiss uplift calculated through a deterministic approach based on expert opinions – the other elements of costs, estimation uncertainties and risks and opportunities are calculated on a probabilistic basis for all countries. The estimation uncertainties and the risks and probabilities were not calculated in 2014 by ANDRA]; SKB 2019; ONDRAF/NIRAS 2020). HLW: high level waste; ILW: intermediate level waste. For a color version of this figure, see www.iste.co.uk/percebois/nuclear2.zip*

Finally, this comparison shows that the contingencies (financial provisions for additional costs indicated in yellow and blue on the graph, for estimation uncertainties and risks and opportunities, respectively) are a function of the maturity of the project. These are considered more important in Belgium (conceptual stage, no national policy, no selected site) than for Switzerland (more advanced stage and limited number of sites considered) and Sweden (only one selected site). Finally, it should be noted that in France, the Ministry of Ecology, Sustainable Development, and Energy, by ministerial decree dated January 15, 2016, has set a target cost related to the implementation of long-term management solutions for high-level radioactive waste and long-term average activity at 25 billion euros at the economic conditions of December 31, 2011, with this target cost to be reassessed regularly at the key stages of the Cigéo project.

1.2.3. *Financing the long-term management of radioactive waste*

1.2.3.1. *Introduction*

This second part follows section 1.2.2 (as indicated previously) and focuses on the financing mechanisms related to the management of radioactive waste, and which are necessary in order to cover the costs, whose cost analysis is detailed in the previous section.

1.2.3.1.1. Problems posed by financing

The financing of the storage of radioactive waste has few equivalent areas in the current industrial world. This is due to the exceptional scale of the time factor, and also to the socioeconomic context that surrounds nuclear power generation and the waste it generates. Unlike conventional industrial projects, the significant investments that will have to be made to realize the storage facilities will not give rise to any financial income in return. On the other hand, the temporal aspect of the storage project falls outside the scale of that of traditional industrial projects, given that it can cover one to two centuries. The closest analogy that can be found is that of the coverage of pensions, either by legal distribution mechanisms, or by financial mechanisms such as those of pension funds.

Adequate and efficient financing of the storage in question can be evaluated on the basis of three objectives: (1) to ensure that the amounts are sufficient and available in a timely manner to cover the entire costs; (2) the financing should be distributed equitably among those who produced the waste, while being economically efficient to cover the storage costs; and finally (3) that the environmental and safety objectives of the storage are achieved.

It will become clear that these three criteria are partly contradictory, given the many uncertainties that will inevitably arise over the very long-term. This means that no method of financing is perfect. For example, a levy applied by kWh of electricity certainly gives a sufficient guarantee of generating the necessary funds in the long-term, but future payments could be borne by consumers who have not themselves benefited from past nuclear power generation. The polluter pays principle is now widely accepted as the best approach most effectively achieving these objectives.

In this context of the polluter-pays principle, most countries with a nuclear power program have created a managing agency, later called RWMA (radioactive waste management agency) for all or part of the management of radioactive waste and the financing of the storage of this waste. Such an approach is in line with the "classic triangle", which must ensure as best as possible the absence of conflicts of interest between the RWMA, the waste producers and an independent control authority managed directly by the State, which will be responsible for verifying the existence and provision of services within the proper time limits to cover the costs related to storage. In France, the National Agency for the management of radioactive waste (ANDRA) has the mission of managing radioactive waste and carrying out geological storage. This RWMA is not directly involved in the financing, but receives payments from the three nuclear operators, EDF, Orano (ex-Areva) and the CEA, for its operations. The French supervisory authority for financing is the CNEF, the National Commission for the Evaluation of Financing (dismantling of nuclear installations, spent fuels and radioactive waste).

1.2.3.1.2. Purpose and content

The objective in this case is to outline the contours and the main principles of several different modes of financing applied in the countries of the European Union that have important nuclear power programs. This is a question of presenting the advantages and disadvantages in each case, as well as the risks involved in relation to the polluter pays principle, and to controlling and guaranteeing measures that can be taken to overcome such risks. Each of these countries apply a different approach to financing, but we can distinguish two main types of approaches for how and provisions that will be used to carry out storage in the best conditions of compliance with the three criteria given above:

– advance cash payments on the kWh produced in the country;

– payments *on a pro rata basis* of deferred amounts for expenses and/or waste handed over to the RWMA for storage (and, if under the responsibility of this entity, for conditioning and intermediate storage).

Each of these approaches and mixed approaches adapted according to the type of costs to be covered are subject to different mechanisms depending on the means of collecting funds, the management of the visions for deferred expenses and the shared responsibilities of the producers, the RWMA and the supervisory authority. The advantages and disadvantages of each approach to the formation of assets for financing are described in the first subsection. The second subsection describes the method of calculating the liabilities, as well as the risk-bearing parameters of influence that affect – and complicate – these assessments.

The third subsection describes the practical mechanisms of prudential management to diversify the risks of default in the long-term, the so-called "asset-liability" management, or modern portfolio management.

The final subsection describes the role of the State in its mission of controlling the funds and provisions created for storage activities. A summary of the European Union countries with noteworthy nuclear programs is also provided. The contrasting cases of France, Belgium and Germany are described in greater detail.

1.2.3.2. *The methods of financing – revenue collection (assets)*

The reader will find detailed information in some previous studies (European Commission 1988, 1999; Kunsch and Emmery 1996; Kunsch and Mathieson 2001; IAEA 2020). The chronology of the expenses for carrying out the geological storage of HLW, intermediate-level waste (ILW) and long-lived ILW extends over very long periods, and there will inevitably be significant delays between the time when the revenues are collected and the time when the expenses are incurred. Therefore, there is a very high level of uncertainty as to the exact values of these amounts, resulting in possible additional costs with respect to the initial estimates, even after taking into account the margins of uncertainty and the adjustments made during the project as described in the section relating to "immediate" costs above.

The most robust possible financing methods for most effectively fulfilling the three partially antagonistic criteria described above must be found and implemented, bearing in mind that no approach (or mixed approach) will be perfect in ensuring the sustainability of financing throughout the entire chronology without involving future generations who have not benefited from the production of electricity with which these wastes are associated, and that should intervene in the financing if it proves insufficient. In relation to this last point, for a correct application of the polluter-pays principle, it can first be noted that the collection by the State through taxes or fees on expenses *pro rata*, and therefore by a method of distribution similar to that used for legal pensions in France and Belgium, would certainly be a safe solution, though not necessarily a fair one, as it is based on an assumption of the long-term continuation of a nuclear power program sufficient to cover the entirety of the charges of the past, which will not necessarily hold true. In all methods for financing storage activities, it is thus necessary to provide for the financing of deferred expenses by provisions using methods that may differ by country. We also note that several financing mechanisms can coexist within the same country.

1.2.3.2.1. Payment of cash advances on electricity generation

This method consists of collecting a direct debit (*levy*) based on the kWh produced in the country, which is then deposited into a fund. This fund may be managed by a specialized organization, under the control of the supervisory authority, the RWMA or both. In the European Union, this method of financing is practiced in Spain, Sweden and Finland, though with very large differences based on the type of collection and the sharing of responsibilities. For example, Spain collects a levy on each kWh produced – whether or not it uses nuclear energy – and the fund is managed by the RWMA (Enresa), under the control of the state (as a dependency of the state), while only nuclear power generation is subject to levies in Sweden. These are collected for each nuclear generation facility to raise funds, under the control of the French nuclear safety authority.

The sample must be reassessed periodically. The charge is calculated in a fixed currency on the date of the revision, and it is indexed each year to adjust to the current value (CV). Future costs are discounted at the date of the revision on the basis of a discount rate, which may be the average return on the fund's investments.

The advantages of this method of financing by advance payment at the time of electricity production are the following, as compared to the polluter pays principle:

– Fairness in relation to current electricity consumers is respected (although in the case of Spain, all electricity production participates in payments). The deferred costs are therefore properly covered by those who have benefited from the production of electricity.

– A serious guarantee exists to arrange for financial assets to cover future liabilities, even if a particular electricity producer were to cease activities at the time such activities were being carried out – because it can be assumed that electricity will always remain an essential good – and if it is accepted that the fuel costs can be attributed to the entire electricity production, as in Spain.

For expenses closer in time to R&D and base investments, we will have a significant financial cover without having to resort to borrowing, for which the costs would be added to the total technical costs.

– This approach generally assumes the presence of a fund managed by a specialized organization. The management according to an asset-liability type approach – described in section 1.2.3.4 – allows for future additional costs to be compensated to a certain extent by generating capital gains on assets and financial products.

This approach also comes with certain disadvantages:

– The estimates of the fees are complicated by the uncertainties relating to the costs mentioned earlier, in addition to economic factors: the average inflation of the economy, which can be assumed to be 2% per year according to the ECB's objectives in the euro zone, but also differential inflation on the costs of materials and services, etc. Also, it is difficult to predict the evolution of the financial returns produced by management, and which serve as a basis for forecasting an average discount rate over the long-term. In the event of large increases in the levy required by the excess costs, the polluter pays principle is no longer be respected if these charges are passed on to future generations.

– It is not certain that the accumulated amounts are sufficient, or available in the form of cash, to cover R&D expenses, or even current investment expenses. Indeed, a significant decrease in the value of assets – private or sovereign bonds or stocks, which we will discuss later – could make it difficult to create assets without prohibitive losses.

– This approach assumes the sustainability of the nuclear power producers in charge of the collection, except to accept that a transfer is possible to all electricity production in general, including conventional production (see the case of Spain above). But this condition will not necessarily be fulfilled, since several European countries, including Belgium and Germany (see section 1.2.3.5.3), have decided to discontinue their nuclear power production, and future producers could refuse this transfer of charges.

1.2.3.2.2. Collection of deferred amounts

Pro rata collection of expenditures

This case is the simplest one. The producer pays annually on the basis of annual or multi-annual budget forecasts for expenses as they arise. The current costs in the year of budgeting are used.

This method of financing has the great advantage of avoiding advance payments to cover future expenses, for which the estimates are subject to great uncertainty, while with this approach the economic context is known, allowing for short-term adjustments of payments.

It has the disadvantage of not respecting (as-is, without provisioning the fuel charges) the polluter pays principle, since most of the storage costs are deferred. This simple approach is only well suited to covering current expenses that occur within a short time span, such as for interim storage, R&D costs or expenses upstream of the actual storage. For expenses related to storage operations, it is therefore necessary to assume a continuity of the electronic program, and the producers themselves, throughout the duration of storage in order to be in the same conditions as a distribution system, such as that of legal pensions. Since this continuity cannot be guaranteed, it will be necessary to combine this simple financing modality with a provisioning of future expenses by the producers, as is the case in France with Orano, EDF and CEA.

Pro rata contributions for waste placed in intermediate storage or long-term storage

This approach is more complex, given that it can take several forms, depending either on the moment when the waste is placed in intermediate storage, or stored directly. Different methods of calculating the fees are possible, depending on whether the volume alone is taken into account, or in addition to the physical characteristics of the waste, according to a so-called multi-attribute approach that makes it possible to better respect the equity between waste producers. This approach was chosen in Belgium by the ONDRAF/NIRAS RWMA, which applies a price per unit of volume differentiated according to the type and activity of the waste to be stored when receiving the waste for storage. Due to the delayed payments over time, such as at the time of storage of the waste after packaging, producers continue to be required to make provisions for future expenses, as in the previous case.

The advantage of deferred payments (for which it will be possible to assume a periodic adaptation of the provisions made by the producers) is that the economic conditions will be better known at the time of payments, which will be closer in time upstream of the actual storage.

The disadvantages are as follows:

– In the event of late payment (e.g. payments solely at the time of storage), the polluter pays principle is not better respected than in the previous case, since it is again assumed that the producer will be sustainable until the time of payments. However, these will occur before the start of storage, which partly limits the risk associated with the cessation of activity.

– Loans may have to be taken out, the costs of which would be added to the technical costs, to finance the expenses upstream of the storage, since the only revenues would be those of the waste already received. However, cash advances for

R&D work and initial investments may be imposed on producers if necessary, as is legally provided for in Belgium (see section 1.2.3.5.2).

1.2.3.2.3. Conclusion on the income

The constitution of a fund managed and funded by fees on the kWh produced best respects the polluter pays principle, if it takes the form of a distribution system, in which nuclear power generation can continue to ensure in the future the financing of all storage costs, without time limits. However, this future existence of a share of nuclear power generation that forms a sufficient proportion of the energy supply cannot be guaranteed with certainty, as shown by the laws on the termination of nuclear power adopted in several countries in Europe, including Belgium and Germany (see section 1.2.2.5).

All other solutions to the *pro rata* expenses or reception of waste, if no provision is provided for by the operators, are not truly satisfactory since they constitute withdrawals delayed in time compared to the production of kWh, and therefore of the waste to be stored. They must therefore necessarily be associated with provision mechanisms recorded in the accounting balance sheets, but also providing for the consideration of available assets (dedicated assets: shares, bonds, real estate, financial guarantees, etc.) that are sufficiently liquid at the time of their realization. With this addition, they are satisfactory in terms of intergenerational equity, but remain associated with great uncertainties given the temporal extent of the storage.

1.2.3.3. *Methods for calculating liabilities*

As shown previously, for all modes of financing that respect the polluter pays principle, it is necessary to provide mechanisms for making provisions for deferred storage charges, which in all countries with a nuclear power program constitute the largest part of the amounts to be provided (*future liabilities*) (OECD/NEA 1996, 2013; Lévêque 2013). The proper calculation of these charges is therefore crucial to follow the polluter pays principle as closely as possible. We will now examine this calculation, as well as its influencing factors.

The calculation of future expenses can be done either in a constant currency in the year of the evaluation, or in current currency of the year of their occurrence. In most cases, we will choose the calculation in constant currency, for at least two reasons. First of all, the calculation in current currency, which therefore takes into account an estimated inflation rate, gives rise to a very large increase in the amounts spread out over time. They therefore become difficult to interpret compared to the estimates of the so-called "immediate" costs (or overnight costs) obtained by the methodology described in section 1.2.1. For example, considering an inflation rate of 2% which is the average target pursued by the European Central Bank in the euro zone, the amounts are to be multiplied by $7.24 = 1.02^{100}$ over a span of 100 years.

This average target also does not take into account the differences in the evolution of the costs of certain works, basic materials, or equipment, which could differ greatly from the average inflation rate. On the other hand, the consideration for inflation presupposes a specific knowledge of the chronology of costs, which is rarely the case. The calculation in constant currency can in turn be made according to two different methods: in CV or NPV. The practices differ according to the financing methods described above.

1.2.3.3.1. The CV method

This method consists of equating the future burden to the immediate costs increased year by year, allowing inflation to be taken into account, but without taking into account the chronology. This is very advantageous for the formation of provisions for future expenses for several reasons:

– Due to the fact that we can ignore the assumptions relating to the times when the expenses will take place, we can make provisions as soon as possible as soon as the expenses are identified and known in current currency of the current year.

– Over time, the accounting provisions or dedicated assets recorded in the books of producers can be adapted on an annual or multi-annual basis, depending on periodic revaluations and taking into account the various inflation factors at the time of the revaluation.

– Provisions can be built up relatively quickly, which represents an advantage for producers if they can deduct them for tax purposes, as is the case in most OECD countries.

– In the case of the management of provisions in a dedicated fund, we can benefit at the earliest time possible from interest income, which will ensure additional coverage of expenses. Adaptations of the sampling can also be provided more quickly.

This CV approach therefore offers security in terms of the availability and adequacy of the amounts provided with respect to remote costs, and because these are collected as early as possible, they reduce the involvement of future generations in covering the costs of storage. However, it presupposes the sustainability of the projects, which probably also implies that of the nuclear power program that generated the waste. Although it can be assumed that electricity will continue to be a necessary good, there is still a risk that the periodic adaptation of provisions will no longer be possible due to the disappearance of this program or the diversion of provisions, whether managed or not, for other purposes.

1.2.3.3.2. The NPV method

This NPV methodology is more complex. It consists of updating all future expenses and revenues on a single reference date using the updating technique explained in Chapter 2 of Volume 1. That is why this approach is also commonly referred to as the NPV method.

The NPV therefore makes it possible to calculate (and periodically revise) the amounts to be calculated annually to cover future expenses assessed within the framework of one or more scenarios with different timelines, whether for a method of financing by a levy on kWh or on waste handed over to the RWMA or put in storage. This then makes it possible to consider the accounting provisions or managed in a fund as the "savings" of the storage system, which makes it possible to adapt them over time, but in practice only during the period of electricity production and waste management and relatively later than in the CV approach.

This method indeed has two major disadvantages compared to the VC method:

– It accumulates funds much more slowly than the VC, and, in both cases of levies on kWh and on waste, the funds collected could be debited, which would require the use of borrowing or a taxation mechanism to compensate for the deficit.

– It is understood directly from the definition of this method of calculation that it is by far much more complex than the previous one, and that it involves additional influencing factors related to the scenario(s) adopted for the storage, i.e. the chronology itself, and also the choice of one or more discount rates that would making it possible to calculate the value at the present reference date, such as January 1, 2021. This last choice in particular potentially gives rise to issues at the limits of ethical philosophy, which we will not be able to address here. We will find many references to this in the literature, (OECD/NEA 1994; Loisel et al. 2004; Gollier 2005; Kunsch et al. 2008; Lévêque 2013; Doan 2017). The choice of the discount rate is critical, since with even a low positive rate, expenses occurring much later on are "erased" in the economic calculation. It is therefore an essential and limiting parameter for the calculation of provisions for future storage amounts. Many authors thus argue for a decreasing rate over time to avoid the phenomenon of erasure in the distant future. Gollier (2005) recommends a nominal rate of 5% per year for costs incurred before 30 years and 2% for costs incurred beyond that time.

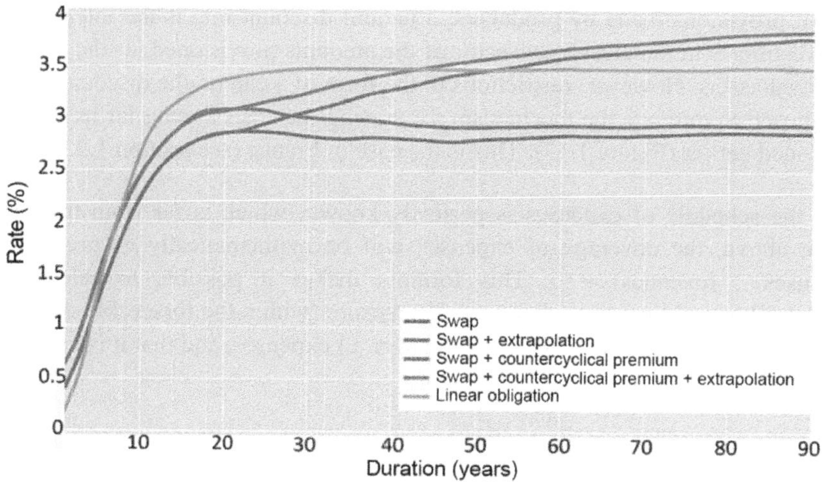

Figure 1.17. *Curves for linear bonds (OLOS) and swap contracts (with several extrapolation approaches up to 90 years, while the maximum duration of a sovereign bond (OLO or OAT) is 30 years), which can be chosen as nominal discount rates as risk-free rates (with the situation as of 2013). For a color version of this figure, see www.iste.co.uk/percebois/nuclear2.zip*

Within a strictly financial framework (which assumes that financial markets comparable to ours will exist throughout the storage period, among other assumptions), the discount rates are primarily linked to the financial performance of investments in the dedicated fund, which can be supplied using mechanisms such as fees on kWh and generating interest at the real rate of return (excluding inflation) (Kunsch et al. 2008; Lévêque 2013). If the fund is managed by the RWMA, one prudent option is to choose a risk-free rate, in other words one that provides a present value of future expenses excluding investment risks. The rates of sovereign bonds of nations (OAT in France, OLO in Belgium, Bund in Germany, etc.) in Figure 1.17 can be considered as risk-free rates compared to equities as long as we neglect the credit risks, in other words a default by the borrower, in this case the nations, for their financing[14]. An asset-liability management (ALM), also explained below, would make it possible to generate added value compared to the risk-free rate of the assets managed in the portfolio to better cover the risk of insufficiency. In the

14 Credit risks can be found in the use of swap rates, exchanges of variable rates for fixed rates with a borrower considered as less risky. As a result of this risk, the *swap* rates which are established by mutual exchanges between banks (swaps) are lower than the rates of sovereign bonds, the Belgian OLOs, as given in Figure 1.17.

case of provisions made by producers, a natural discount rate is the internal return on investments in the electricity sector of the amounts provisioned, or the returns on dedicated assets. However, restrictions in the form of a cap on the discount rate may be imposed to mitigate the mechanism for erasing the costs arising far in the future, mentioned earlier (Figure 1.17). This is the case in France (see section 1.2.3.5.1).

If the schedule of expenses is perfectly known, which is far from the case as shown above, the coverage of expenses will be mathematically ensured if VPN (expenses – revenues) = 0. This formula makes it possible to calculate the withdrawal at any time in such a way as to ensure (within the forseeable future) that the amounts in the fund are sufficient to cover all expenses, and that it is empty once the storage site is closed.

Calculation of a direct debit by the NPV

In order to illustrate the discount method and the calculation of the revenues to be placed in a long-term managed fund to cover the costs, we will choose the method of financing by rates per m^3 of waste delivered to the RWMA at the time of storage (a calculation relating to financing by fees per kWh is based on the same principles averaging the earlier date of collection).

We will look at time t_0 for a reassessment on this date of the tariff V [€/m^3], which will be assumed to be constant in constant currency over all years t_i with i = 0, 1, 2, etc., M where t_M is the year of closure of the storage facility. We use the following definitions for the calculation:

– $F(t_i)$ [€] is the balance of the fund at time t_i;

– $E(t_i)$ [m^3/year] represents the volumes of waste giving rise to payments of fee V at time t_i;

– r [%/year] is the real discount rate in currency. It is assumed to be equal to the actual annual average return of the fund;

– $D(t_i)$ [€] represents the annual disbursements.

The evolution of the fund year by year is calculated by the following recurrence formula:

$$F(t_i) = F(t_{i-1})\,(1 + r) + V\,E(t_i) - D(t_i) \qquad [1.1]$$

The first term of this sum gives the growth of the available fund during the year t_i by the capitalization of the interest at the rate r. The last two terms give for each year the difference between the revenues generated by the rate payments *pro rata* for waste and annual disbursements. For t_M the fund is assumed to be empty, except for a residual amount used for a subsequent control and guarding of the site and/or to keep its long-term memory in order to avoid involuntary intrusions[15].

Subject to this remark, we therefore obtain $F(t_M) = 0$. By placing itself at the end of the period in t_M and then substituting into this formula $F(T_{M-1})$ and so on until going back to t_0, we obtain by recurrence:

$$\frac{F(T_M)}{(1+r)^M} = 0 = F(t_0) + \sum_{i=1}^{M} \frac{V\,E(t_i) - D(t_i)}{(1+r)^i} \qquad [1.2]$$

It is well known that the sum of the discounted receipts plus the balance in t_0 is equal to the discounted sum of disbursements, that is, as previously written: (expenses, t_0) = NPV (income, t_0) + balance at t_0. The constant rate evaluated for t_0 is then given as follows:

$$V = \frac{\sum_{i=1}^{M} D(t_i)\,a^i - F(t_0)}{\sum_{j=1}^{M} E(t_j)\,a^j} \qquad [1.3]$$

where $a = {}^{1}/_{1+r}$ is the discount coefficient at the discount rate R.

An example of calculating the evolution of the storage financing fund on the basis of tariffs at the reception of LL-LLW, LL-ILW and HLW waste is shown in Figure 1.18. The complication that arises when multiple fees are used rather than just one is easily taken into account in the calculations.

This modeling can be generalized to other cases, for example, for calculations of current currency levies including variable inflation assumptions based on long-term projections of the markets, taking into account taxation on income from stocks and bonds, or sensitivity analyses relating to discount rates decreasing over time, to better take into account disbursements very distant over time, etc.

This technique also takes on its full value in the asset-liability valuations described below, for which the goal is to lead to a better coverage of the financial risks of storage by defining an efficient asset allocation (AA). For this purpose, simulations known as Monte Carlo simulations are used, which have been discussed

15 To this end, it will be possible to continue indefinitely, producing interests without starting the initial capital, a financial situation that can be found in instances such as certain cultural foundations, some of which have existed for decades or even centuries. To do this, a residual amount is reserved in the fund $P = I/r$, which for each year will generate the amount of interest I equal to the average operating costs I by investment at the average rate of return r obtained by the fund.

in section 1.2.2.5.6. For this, a large number of scenarios are generated by random draws on the returns that are expected but whose distribution is uncertain given their volatility. On this basis, an amount of direct debits will be chosen allowing a high level of confidence for the coverage of the charges over the entire span of the calculation period. An example of such a calculation based on ALM and a stochastic analysis of the evolution of the fund's returns according to an equity/bond AA are given in the reference (Kunsch et al. 2008).

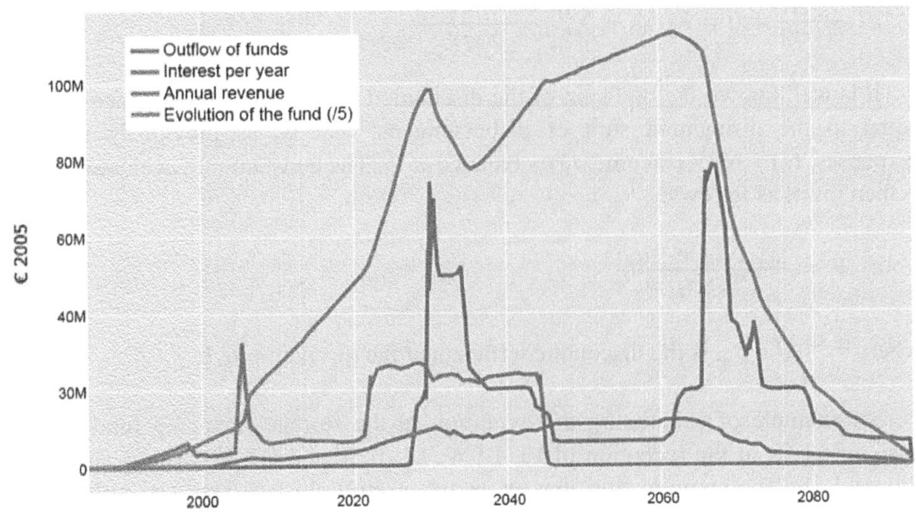

Figure 1.18. *Deterministic modeling of the evolution of the storage fund (HLW, LL-ILW, LL-LLW) generating interest for the calculation of tariffs (the costs in constant euros 2005 are given purely by way of illustration, the real interest rate is 2%/year). For a color version of this figure, see www.iste.co.uk/percebois/nuclear2.zip*

1.2.3.3.3. Current value or NPV

The influencing factors in the expense calculations must be carefully examined when developing a calculation methodology in relation to legal constraints and the sharing of responsibilities between the actors of the classic triangle regarding the management of the provision of funds.

The CV approach should be preferred if the provisions are primarily noted in the books of the producers, which can be partly reinvested in current projects, and if it can be assumed that the nuclear power program and the responsibility of the producers who ensured this production are sustainable. The main influencing factors are unforeseen events and hazards affecting storage costs. Adjustment and reassessments are easier to operate with this approach because there are no purely financial uncertainties related to the adequacy of funds and the storage schedule, as well as inflation factors, taking into account this sustainability of producer responsibility. The amounts provided will be established shortly after the completion of the project, without the contribution of financial returns in a managed fund. This largely avoids the use of borrowing, advance calls for funds, or state intervention through taxes or a tax for R&D and initial investment costs.

However, the NPV approach is preferable in the case of the creation of a managed fund, or dedicated assets in return for accounting provisions – two frequently occurring cases. In the case of a fund, the collection of financial resources is slower than in the CV approach, although it may benefit from income from interests. This may make it necessary to obtain loans, or even to tax for the first investment phase of the project. As for dedicated assets, the cost estimates are very sensitive, primarily to the timing of the collection of funds and the realization of storage, as well as the choice of discount rate(s), which can possibly be made variable over time to avoid the erasure of discounted costs in the far future. If we choose a low discount rate lower than the average return of the managed fund, such as a risk-free rate swap, it will be possible to form a reserve to compensate to a certain extent for the additional costs of the project in the longer-term.

1.2.3.4. *Asset-liability management of managed funds*

The methods for forming assets and calculating expenses have been described in the two previous subsections. Whatever the different approaches chosen by the countries, mechanisms must be put in place to cover the financial risks as thoroughly as possible and ensure that the funds established in accordance with these methodologies remain available. Financial security is defined as the necessity to ensure the availability of liquid funds at any time they are necessary to carry out storage activities with a minimized degree of risk. The financial risk that comes into play in this definition is thus a risk to liquidity and yield, also known as a rate risk, that jeopardizes the sufficiency of funds under unfavorable market conditions.

The most commonly used mechanism to ensures the financial security of funds to best cover these risks is referred to as ALM, briefly described in Box 1.2. This technique falls under the modern portfolio management technique introduced by H. Markowitz in 1952 (Broquet and Hübner 2019). It is based on a technique for setting up diversified asset investments to be placed in the management portfolio according to the principle of sound judgement that cautions against "putting all our eggs in one basket". Based on a risk profile of the investor – in this case, the RWMA or the producer – an AA seeks to define an efficient investment portfolio by minimizing volatility, and therefore the level of risk represented by the standard deviation of the return, among all the portfolios with the same average return (see Box 1.2). This technique is currently used in France for the management of dedicated assets used to cover the provisions of funds recorded in the balance sheets.

Asset-liability management (ALM) aims to optimize at any time, throughout the duration of a project, the coverage of the liabilities by assets which are liquid, made available promptly, and in sufficient quantities, so as to minimize all additional financial costs, such as the use of borrowing, or state interventions for last-resort financing.

The financial risks of liquidity at the time of use of the funds and the risks of return of the funds are covered under the definition of a distribution of assets between investment products according to the asset allocation (AA) technique. This technique considers the proportions of different assets: stocks, government bonds or corporate bonds, liquid investments and real estate. The volatility of the portfolio is reduced by the choice of assets that are not correlated or negatively correlated with each other. We thus benefit from the risk premium of higher yielding assets while limiting the risk of volatility on the markets according to the pairing of (μ, σ), where μ [%/year] is the average return on the portfolio and σ [%/year] is its standard deviation measuring this volatility.

For long-term expenses, the AR should ideally be oriented toward a relatively large proportion of good quality shares, but whose prices show little or no correlation to obtain an efficient portfolio, which makes it possible to generate growth in the fund's assets and interest income, giving a better coverage of long-term expenses and therefore to reduce the risk of insufficiency. An AA in favor of more stocks is based on the observation that investments in cash or bonds have lower cumulative returns over the long-term. However, equity investments are clearly characterized by a higher return, though also by a greater level of risk, due to their higher volatility. For this reason, equity investments are not liquid and should not be considered for the short-term. Most countries want to maintain a low-risk profile for the ALM, and they limit the AA to more conservative investments such as government bonds or liquidity term investments whose returns are significantly lower but whose volatility is reduced.

Box 1.2. *Asset-liability management (ALM)*

As part of an AA, government bonds or corporate bonds of good quality (with a AAA or AA rating) constitute investments of lower risk with returns close to a risk-free rate, but they do not completely protect against liquidity risk, because in a period of growth in interest rates the value of bonds decreases, resulting in a loss on sale before maturity. Only liquid investments will by definition allow for the liquidity risk to be addressed, but even investments in term accounts or government bonds are subject to uncertainties related to the evolution of the yield curve.

It should be noted that in 2020, this yield curve as a function of duration (here referring to the duration for bonds) has negative values for short durations and is close to zero for durations of 20 years or more. It is very difficult to predict when this curve will resume a "normal" pace that is positive and increasing with the duration of the placement. A successful AA, and one most often evolving according to the deadlines of expenses, is therefore a delicate exercise, because it is also subject to institutional constraints within the framework of prudential and most often justifiable management, because any speculation must be avoided on the part of the fund's managers, the trust companies (financial managers of an investment fund) which are subject to constant controls and regular audits by the beneficiaries of the fund, producers, RWMA and the government.

The ALM/AA approach will make it possible to take advantage of a risk premium obtained by investments with higher returns than liquid returns deemed "risk-free" (the swap rates are used here as risk-free rates for discount calculations). In this way, it will be possible to create added value to the fund in order to best prevent it from failing over the long-term. To define this more efficient AA, we will compare different investment policies using a Monte Carlo analysis, focusing on investment rates and costs: first a policy based on liquidity investments at the risk-free rate which was used for the calculation of the NPV, and second, riskier investments comprising various holdings of bonds and stocks (ONDRAF/NIRAS 2015).

In conclusion, the application of ALM management leading to an efficient AA is important to cover the financial liquidity and interest rate risks as effectively as possible in the case of provisions managed in a fund, or dedicated assets as a counterparty to provisions. However, this only constitutes an ideal case, because it remains in most cases inapplicable as it is, due to the constraints imposed by most governments on the level of risk permissible for AA (i.e. a "low-risk" investment profile for asset portfolio management), which limits it most of the time to investments in government bonds and liquid investments at rates close to the risk-free rate.

1.2.3.5. *The role of governments and the mechanisms for controlling provisions*

This last section analyzes the role of governments and the financial supervisory authority in the countries of the European Union (European Commission 2011).

To ensure the nature of this section remains that of a synthesis, only the cases of France, Belgium and Germany are presented. These same principles apply in most cases to future charges attributed to the dismantling of nuclear reactors.

1.2.3.5.1. Financing storage costs in France

In France, the existence of a strong presence of governmental agencies in all nuclear activities has made it possible to decentralize the responsibility for raising funds to the three nuclear operators: the two operators, Orano (ex-Areva) and EDF, and the CEA for nuclear R&D. The principle of this decentralization, which remains under the control of the government, is the strong assurance of a sustainability of electricity production and no doubt nuclear power as well, which remains and will remain a necessary and also profitable activity in the long-term in this case (in 2020, about 70% of electricity is produced by nuclear reactors). There is little fear of insufficient long-term financing, with the government appearing here as the guarantor of the adequacy of the provisions and as a last resort for financing. The method of financing is the *pro rata* collection of deferred amounts of expenses, which will be paid to ANDRA (RWMA) to carry out the storage. To respect the polluter pays principle, the formation of a funding supply within the chief of the producers is necessary, with the counterpart of dedicated assets.

The basis for this method of financing is the law of June 28, 2006, establishing the environmental code, and specified by the decree of February 23, 2007 and the decree of March 21, 2007 (with the latter relating to securing financing for nuclear-related expenses), which defines the obligations of the operators of nuclear installations and organizes how they are overseen. The approach to financing long-term nuclear charges (Ministry of Ecological Transition and Solidarity [*Ministère de la Transition écologique et solidaire*]) 2020) respects the polluter pays principle and is formulated according to the below-mentioned principles. It is commonly used in the management of spent fuels and the dismantling of reactors.

Principle of conservative estimation: operators enter provisions in the liabilities of their balance sheet. Any future charge must be estimated conservatively, such as by using margins compared to the best estimate, and also by choosing the discount rate which is capped by the decree of March 21, 2007 relating to securing the financing of nuclear charges modified in 2020[16].

Principle of pre-financing: the provisions must have a counterpart in dedicated assets that will be guaranteed to be available to meet future expenses. These assets must have a sufficient degree of security and liquidity, and the regulation regulates

16 The real value of the ceiling is equal to the non-rounded value representative of the long-term real interest rate expectations, used for the calculation published by the European Insurance and Occupational Pensions Authority of the ultimate term rate applicable on the date considered, plus 150 base points. This ceiling is applicable beginning in the year 2024. Up to that year, the ceiling is equal to the weighted average of 2.3% and this new ceiling.

the composition of the portfolio. The AA according to the ALM portfolio management methodology is also referenced (see the report of the Court of Auditors (2012)). "The law makes it mandatory to secure the financing as soon as the installations are put into operation or as soon as a new fuel is introduced into the reactor core". The explanatory diagram of the financing mechanism is given in Figure 1.19. As can be seen in this diagram, the discount mechanism is used for the calculation of the dedicated assets needed today. The choice of a discount rate is therefore very important as part of a prudential approach for the calculation of these assets (including French or foreign stocks outside the scope of the operators, government or corporate bonds, real estate, etc.). The discount rates are regulated by the legal provisions of the decree of February 23, 2007 and its modifications relating to the securing of the financing of nuclear expenses mentioned above.

Principle of confined internal funds: dedicated assets must be treated as internal funds to the operator, separated from the rest of the balance sheet. Their use is strictly limited to the coverage of long-term nuclear charges. This principle is an application of the European directive EUR 2011/70[17] containing the requirement to create dedicated funds (known as *ring-fenced* funds) for long-term nuclear loads, including for the dismantling of installations (European Commission 2011).

Total and infinite responsibilities of the operators and the parent company: this responsibility extends to the estimates of future expenses, the constitution of dedicated assets updated according to the performance evaluations of the asset portfolio, etc., and the realization of operations when the time comes. In the event of the default or insolvency of the operator, the obligations are assumed by the parent company. This full and complete responsibility of the operator and its ownership structure provides an effective response to the concern to ensure the sustainability and the adequacy of the financing over the entire span of the execution of the storage (or dismantling).

The strict control of these obligations (provision of triennial reports by the three contractors, annual updating of this report, obligation to provide information, accounting audits by the Court of Auditors (2012), etc.) is accompanied by the power to sanction and is ensured by the administrative authorities, in this case the CNEF, created by the law of 2006 under the aegis of the Ministries of Economy and Energy.

17 The EU Member States must put in place a robust and effective framework as well as an independent regulatory authority; they must also ensure that the financing mechanisms that are chosen guarantee the availability of the funds.

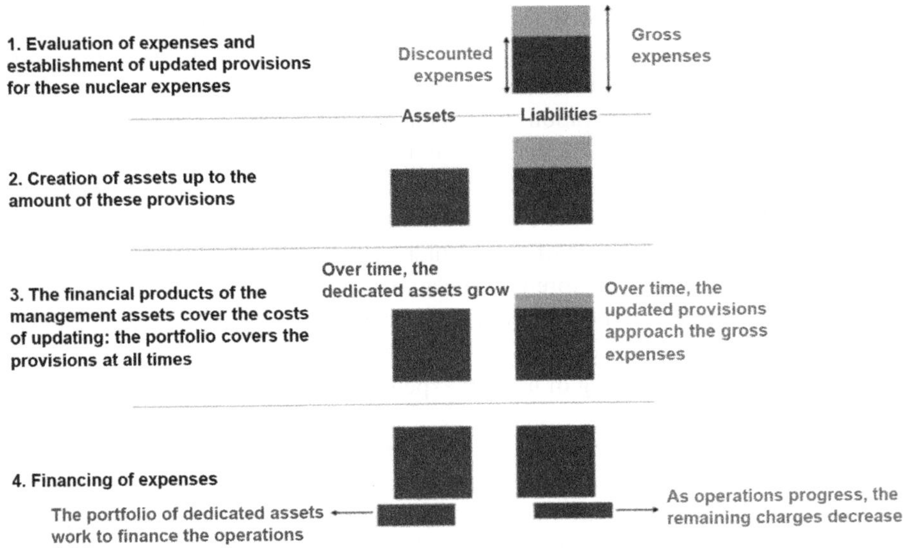

Figure 1.19. *Explanatory diagram of the financing mechanism borrowed from Ministère de la Transition écologique et solidaire (2020), concerning the assessment of expenses to define the amount of provisions made in the liabilities of the balance sheet and the calculation of dedicated assets. For a color version of this figure, see www.iste.co.uk/percebois/nuclear2.zip*

1.2.3.5.2. Financing storage costs in France

In Belgium, the sustainability of nuclear power activities and producers does not exist due to the 2003 nuclear removal law, which provides for the shutdown of the seven Belgian reactors between 2023 and 2025[18].

ONDRAF/NIRAS (National Organization for the Management of Radioactive Waste and Enriched Fissile Materials; NIRAS is the Dutch translation) was created as a RWMA by the law of August 8, 1980. Its capacities and duties were established by royal decree in 1981, they involve the comprehensive management of radioactive waste. The 1981 RA has been subject to successive modifications; specifically, provisions including the method for calculating the fees applied to wastes received by the organization were added in 2014. The organization is supervised by the ministers overseeing the economy and energy.

18 The Belgian government announced in March 2022 the extension of the service life of Doel-4 and Tihange-3 beyond 2025 and for a period of 10 years.

A mixed financing system has been set up combining the creation of a long-term fund fed by tariffed levies at the reception of waste by the RWMA and the establishment of provisions for the operators of nuclear power producers for the management of spent fuels and decommissioning waste from the seven nuclear reactors in the country.

ONDRAF/NIRAS collects the funds raised as fees at the time of receipt of the packaged waste for storage and subsequent storage: the rates for warehousing and storage are applied to each category of waste. The funds are invested in the long-term fund managed by ONDRAF/NIRAS. It should be noted that, for non-reprocessed spent fuels, an initial interim storage is provided directly by the producers (wet or dry storage on the sites of the power plants) before reprocessing or conditioning the assemblies.

With regard to the management of spent fuels – whether or not they are reprocessed – in their entirety (on-site storage, reprocessing or conditioning assemblies, warehousing and storage), the mechanism for financing operations is decentralized with respect to the nuclear electricity producers who are required to make provisions at the time of unloading the spent fuel from the reactor (coverage of the back-end of the nuclear fuel cycle). This field is regulated by the law of April 11, 2003 on the provisions established for the dismantling of nuclear power plants and for the management of fissile materials irradiated in these power plants. These provisions, as is the case with decommissioning provisions, are outsourced within the Nuclear Provision Company (French: Société de provisions nucléaires [SPN]). The control of the SPN is exercised by the "Commission on nuclear provisions", created at the same time as the SPN by the law of April 2003. Three-year revisions of the calculation of provisions are called for. The provisions can be used for the producer's own investments up to a maximum of 75%; the remaining 25% are in the form of available fund, subject to compliance with solvency ratios. These provisions have been the subject of proposals for modifications by the commission of nuclear provisions in order to make their management more prudential.

1.2.3.5.3. Financing storage costs in France

Germany, like Belgium, is carrying out a complete removal of nuclear power in 2022, a decision taken by the federal government (Bundesregierung) in 2011, following the Fukushima accident in Japan. The country's atomic law (Atomgesetz) regulates the management of radioactive waste and the dismantling of power plants. Before the modification of the atomic law in 2011 to organize the removal of nuclear power, the financing mechanisms were comparable to that of France, with pro rata payments of expenses but the formation of provisions for the nuclear power plant operators. In this approach, funds were taken from the operators to cover the future costs of R&D, construction and operation of the storage. The operators – of which there are currently four: E.ON, EnBW (Bavaria), RWE and Vattenfall (Sweden) – kept the full responsibility for the provisions, the amounts of which they had to justify to the accounting clerks or chartered accountants, with the obligation to have sufficient liquid amounts to respond to calls for funds.

The same principles were applied to the dismantling of the facilities. The complete phase-out of nuclear power planned for 2022 led the operators to negotiate an agreement with the federal government that would allow them to limit their very significant operating losses due to the shutdown of the reactors[19]. In June 2017, an agreement was signed with the Minister of the Economy and Energy which, in addition to compensation payments to the four operators, provides for the creation of a state fund to cover all waste management costs including packaging, temporary warehousing and storage. The operators remain responsible solely for the dismantling and conditioning of radioactive waste. The execution and financing of the intermediate storage and the storage of packaged waste are the responsibility of the federal government under this agreement. This transfer of responsibility implies a balance of any account with the transfer of the financial means for warehousing and storage to the newly created fund.

These amounts represented 24 billion euros as at the date of the agreement. However, there is a clear breach of the polluter pays principle in that this transfer of responsibility to the government implies that it will have to fully act in its role as financier of last resort if the financial resources transferred to the fund prove insufficient to cover the entirety of the storage operations.

1.2.3.5.4. Conclusion on provisions

In general, in all countries that have significant nuclear programs, the government remains the most important player in ensuring the best possible implementation of the polluter pays principle. It must take care to ensure independent control of the producers and the RWMA for the constitution and proper maintenance of provisions giving guarantees of availability and sufficient liquidity, and it must constantly be able to compensate for the deficiencies of these provisions to cover future expenses, since it is clear that no system of control, regular audits and financial guarantees will ever be able to fully cover the risks of failures and deficiencies, due to the extensive schedules and the multiple economic and financial uncertainties discussed in this part of the chapter.

19 Agreement signed between the Federal Republic and the energy producing companies concerning the financing of the nuclear phase-out; available at: https://www.bmwi.de/Redaktion/DE/Pressemitteilungen/2017/20170626-vertrag-brd-energieversorgungsunternehmen-finanzierung-des-kernausstiegs.html.

Other institutional or non-institutional risks exist that are not taken into account here. One example of this is fraud on the part of specialized fund managers, or diversions of funds by the government to cover other expenses or commitments, or other budgetary objectives other than that of storage. It is also the role of governments to commit to respecting these financing commitments, as was the case with the United States, which enshrined in its constitution the total immunity of the funds established for the Waste Disposal Trust Fund fed by a kWh tax paid to the Department of Energy (DOE).

1.2.4. *Conclusion on the costs and financing of waste storage*

The costs of storage in deep geological layers as a long-term management strategy for radioactive waste can be established, and the economic and financial techniques to ensure their coverage are currently in place.

Of course, many uncertainties exist and still need to be resolved or mitigated, both from a scientific and technological as well as an economic and financial point of view. However, this does not prevent progress toward the implementation of the proposed solution in application of the precautionary principle, according to which:

> In the event of a risk of serious or irreversible damage, the lack of absolute scientific certainty should not be used as a pretext to postpone the adoption of effective measures seeking to prevent environmental degradation. (Rio Declaration 1992)

In addition, the more concrete a project becomes, a significant number of uncertainties disappear or become less relevant to a greater extent. In economic and financial matters in particular, the transition to the execution phase can significantly reduce uncertainties by materializing costs (knowledge of investments) and setting financial horizons (realization and exploitation periods).

It is first and foremost the responsibility of the producers of radioactive waste to ensure the long-term management coverage of the waste they produce. However, the long periods of time involved and the sovereign decisions taken by the States may lead them to have to provide for part of these costs.

Thus, the costs of long-term waste management can be covered by electricity producers; however, this does not exclude the transfer of part of these costs to the government, the guarantor of last resort. The magnitude of this transfer is an indication of the degree of real application of the polluter pays principle.

1.3. References

AACE (2008). Risk analysis and contingency determination using range estimating. AACE International Recommended Practice No. 41R-08.

AACE (2019). Cost engineering terminology. AACE International Recommended Practice No. 10S-90.

ANDRA (2014). Évaluation des coûts afférents à la mise en oeuvre des solutions de gestion à long terme des déchets radioactifs de haute et moyenne activité à vie longue - Proposition de l'Andra Tome 1 Chiffrage de Cigéo en phase esquisse : synthèse [Online]. Available at: https://www.andra.fr/sites/default/files/2018-02/cout-cigeo-t1.pdf.

Broquet, C. and Hübner, G. (2019). *La gestion de portefeuille. Instruments, stratégie et performance*. De Boeck Supérieur, Louvain-La-Neuve, France.

Cour des comptes (2012). Les coûts de la filière électronucléaire. Report, Cour des comptes.

Doan, P. (2017). Prise en compte économique du long terme dans les choix énergétiques relatifs à la gestion des déchets radioactifs. PhD Thesis, Université Paris-Dauphine, France.

European Commission (1988). Coûts et modes de financement de l'évacuation géologique des déchets radioactifs. Report EUR 11837 FR, Direction générale Science, recherche et développement.

European Commission (1999). Schemes for financing radioactive waste storage and disposal. Report EUR 18185 EN, Environment Directorate-General, European Commission.

European Commission (2011). Radioactive waste and spent fuel management directive. Report, European Commission, Brussels, Belgium.

Flyvbjerg, B. (2014). What you should know about megaprojects, and why: An overview. *Project Management Journal*, 45(2), 6–19.

Flyvbjerg, B. and COWI (2004). Procedures for dealing with estimating uncertainty in transport planning. Guidance Document, UK Department for Transport, London, UK.

Freeze, G.A., Stein, E., Brady, P.V. (2019). Post-closure performance assessment for deep borehole disposal of Cs/Sr capsules. *Energies*, 12, 1980 [Online]. Available at: https://doi.org/10.3390/en12101980.

GAO (2020). Cost estimate and assessment guide, best practices for developing and managing program costs. Report, GAO-20-195G.

Gollier, C. (2005). Quel taux d'actualisation pour quel avenir ? *Revue française d'économie*, 19(4), 59–81.

Havlíček, L. (2008). Nuclear fuel cycle evaluation and real options. *Acta Polytechnica*, 48(3) [Online]. Available at: https://doi.org/10.14311/991.

IAEA (2018). Enhancing benefits of nuclear energy technology innovation through cooperation among countries. Final report, INPRO Collaborative Project SYNERGIES.

IAEA (2019). Management of spent fuel from nuclear power reactors, learning from the past, enabling the future. In *IAEA International Conference on the Management of Spent Fuel from Nuclear Power Reactors*, Vienna, Austria.

IAEA (2020). Costing methods and funding schemes for radioactive waste disposal programmes. Report, IAEA, Vienna, Austria.

Jain, S.A. (2012). Valuing modular nuclear power plants in finite time decision horizon. *Energy Economics*, 36, 625–636.

Kim, S., Ko, W., Youn, S., Gao, R. (2015). Nuclear fuel cycle cost estimation and sensitivity analysis of unit costs on the basis of an equilibrium model. *Nuclear Engineering and Technology*, 47(3), 306–314.

Kunsch, P. and Emmery, D. (1996). A new technique for covering the financial risk of a waste management program. In *Waste Management Conference*, Tucson, AZ, USA.

Kunsch, P. and Mathieson, J. (2001). Financing radioactive waste storage and disposal. A study for the European Commission. In *International Conference on Radioactive Waste Management and Environmental Remediation*, Bruges, Belgium.

Kunsch, P., Ruttiens, A., Chevalier, A. (2008). A methodology using option pricing to determine a suitable discount rate in environmental management. *European Journal of Operational Research*, 185(3), 1674–1679.

Lauferts, U. (2012). Decisioneering in nuclear energy systems: A real option view. Doctoral thesis, Delft University of Technology, The Netherlands.

Lévêque, F. (2013). La méthode du coût actualisé. In *Nucléaire On/Off ? Analyse économique d'un pari. Partie 1 : Les coûts du nucléaire : repères et incertitudes, comment additionner les coûts ?* Dunod, Paris, France.

Loisel, J.-P., Arnould, J., ANDRA, Barré, B. (2004). *Y a-t-il une éthique de la gestion des déchets radioactifs ?* Vuibert, Paris, France.

Ministère de la transition écologique et solidaire (2020). Coût et financement des charges nucléaires de long terme. Report, Ministère de la transition écologique et solidaire, Paris, France.

NASA (2015). *NASA Cost Estimating Handbook*. Version 4.0. NASA, Washington, D.C., USA [Online]. Available at: https://www.nasa.gov/content/cost-estimating-handbook.

NNWI (2020). The failings of levelised cost and the importance of system-level analysis. Report, NNWI, Santa Fe Springs, CA, USA.

OECD/NEA (1994). Environmental and ethical aspects of long-lived radioactive waste disposal. In *Proceeding of an International Workshop Organised by the Nuclear Energy Agency in Co-operation with the Environment Directorate*, Paris, France.

OECD/NEA (1996). Les charges financières futures liées aux activités nucléaires. Report, OECD/NEA, Paris, France.

OECD/NEA (2006). Advanced nuclear fuel cycles and radioactive waste management. Report, OECD/NEA, Paris, France.

OECD/NEA (2011). Potential benefits and impacts of advanced nuclear fuel cycles with actinide partitioning and transmutation. Report, OECD/NEA, Paris, France.

OECD/NEA (2013). The economics of the back end of the nuclear fuel cycle. Report, OECD/NEA, Paris, France.

OECD/NEA (2017). Addressing uncertainties in cost estimates for decommissioning nuclear facilities. Report, OECD/NEA, Paris, France.

OECD/NEA (2021). Strategies and considerations for the back end of the fuel cycle. Report, OECD/NEA, Paris, France.

ONDRAF/NIRAS (2015). Étude ALM des compartiments du fonds à long terme. Report, ONDRAF/NIRAS, Belgium.

ONDRAF/NIRAS (2020). Evaluation of the overnight cost of geological disposal of category B&C waste – Costing 2020. Report, NIROND-TR 2020-01.

Parsons Jr., E.L (1999). Waste management project contingency analysis. Report, Center for Acquisition and Business Excellence, Washington, D.C., USA.

Passerini, S. (2012). Sensitivity analysis and optimization of the nuclear fuel cycle: A systematic approach. PhD Thesis, Massachusetts Institute of Technology, Cambridge, MA, USA.

Piet, S., Bjornard, T., Dixon, B., Hill, R., Matthern, G., Shropshire, D. (2007). Current comparison of advanced nuclear fuel cycles. Report, Idaho National Laboratory, ID, USA.

SKB (2019). Plan 2019 – Costs from and including 2021 for the radioactive residual products from nuclear power. Report, SKB, TR-19-26.

Swissnuclear (2016). Schätzung der Stilllegungskosten der Schweizer Kernanlagen. Kostenstudie 2016 (KS16). Report, Swissnuclear.

Wigeland, R., Taiwo, T., Ludewig, H., Todosow, M., Halsey, W., Gehin, J., Oakley, B. (2014). Nuclear fuel cycle evaluation and screening. Final report, NEA/NSC/R(2015)2.

2

Nuclear Power in the European Electricity Market

Jacques PERCEBOIS and Boris SOLIER
CREDEN, CNRS ART-Dev, Université de Montpellier, France

The liberalization of the electricity sector in Europe was formalized in 1996 with the publication of the first directive, which had been in preparation for about 10 years. Two other directives then followed in 2003 and 2009; the third, better known as the "climate energy package", introduced ambitious objectives for the promotion of renewable energies and energy efficiency.

The liberalization process was then somewhat disrupted by the concern to promote the penetration of renewable energies by using mechanisms that face difficulty complying with the strict logic of the market. The widespread use of guaranteed purchase prices (feed-in tariffs) to develop the production of photovoltaic solar (PV solar) and wind electricity throughout Europe has destabilized the functioning of the electricity market, and the "green wave" has collided with the "market wave", bringing with it perverse effects on the functioning of the spot market for electricity.

It must be said that at the same time, in 2011, the desire by public authorities to reduce the market share of the incumbent (EDF) by concomitantly promoting the penetration of new players on the market, prompted the parliament to require EDF to sell part of the nuclear production at a price that was also regulated, constituting

Nuclear Economy 2,
coordinated by Jacques PERCEBOIS and Nicolas THIOLLIÈRE.

a new exception to the rules of competition by the market, as we will see in section 2.7.

The liberalization of the electricity sector required the Member States to open up both the activity of electricity production and that of electricity supply. However, the networks remain "natural monopolies", which is explained by the existence of high fixed costs, though they must be "regulated" at the end of the law, and the tariff for the use of public electricity networks (TURPE) is an important component of the electricity price (about one-third of the price including VAT for a domestic consumer). In France, the RTE ("*réseau de transport de l'électricité*", in English: Electricity transmission network) is in charge of electrical transmission, and in particular electrical distribution. These two companies are still EDF subsidiaries at the end of 2021, 50% for RTE, and 100% for Enedis, but they must behave like independent network managers and not favor the original operator EDF. In other European countries, these network managers are often private companies that have cut all ties with their parent company. To allow the creation of a true single European electricity market, the European Commission encourages the Member States to develop transnational interconnections, not only for security reasons, but also to facilitate commercial exchanges of electricity and allow for kWh prices for the consumer to converge to a certain extent. This convergence was partly observed for wholesale prices, but not for prices including VAT, given that tax amounts differed depending on the country.

However, choosing the proper balance of electrical power sources, and therefore the means of how electricity is produced, remains the prerogative of national governments under the terms of the treaties in force and each country is free to choose the way in which its electricity will be produced. As a result, the makeup of the electrical power sources differs greatly from one European country to another, due to both geographical and historical causes. France itself, in its recent Multi-Annual Energy Plan (MAEP) project[1] (Ministry of the Ecological Transition and Solidarity 2019), decided to limit the share of nuclear power in electricity production to 50% by 2035, versus 71% in 2019 and 67% in 2020 (RTE 2021), and this decision was not the result of the "invisible hand of the market", but rather the consequence of political choices, which were also disputed by some. A new MAEP will be adopted in 2023.

In France, the price per kWh, set in part by the market, represents about one-third of the price including VAT paid by an end consumer in the domestic

1 Multiannual energy scheduling.

sector, with the other two-thirds being distributed equitably between network access charges and taxes (VAT, CSPE[2], TCFE[3] and CTA[4]).

2.1. The functionality of the wholesale electricity market

The core of the market system is formed by the *spot* market, a wholesale market on which electricity produced by certain sources and that purchased by others are exchanged hour by hour. The wholesale market includes several subdivisions, including the day-ahead market, the intraday market and the futures market. For example, the day-ahead market sets the price per MWh (the basic unit used here is the MWh, or 1,000 kWh) for the following day, 24 h in advance. However, unlike the events seen in England or in California at the beginning of liberalization, this organized wholesale market is not mandatory, because of the lessons that have been learned from the English and Californian experiences where speculation had been strong. It is thus an optional market on which only a limited proportion (20–40%, depending on the country) of the electricity produced is exchanged in practice. A large part of this electricity is in fact exchanged over the counter, which is to say, in a direct bilateral exchange between a producer and a supplier, which may be a subsidiary of this producer. Note that producers are generally also suppliers through interposed subsidiaries, but the reverse is not true. Many suppliers do not produce electricity and buy wholesale electricity that they sell at retail rates. Experience has shown that this spot market tends to become a kind of *adjustment market*, with a number of producers preferring to sell part of their electricity on the futures market of forwards where the price is less volatile. The price on the futures market is set on the day the transaction is made, but the delivery is only made after a time period that can range from a few days to weeks, months or years later. The wholesale market nevertheless plays an important role, since its price serves as a reference point in the contracts signed between suppliers and consumers, and in practice it is taken as a signal for investors. In the event of tensions between supply and demand, prices will go up and producers will be encouraged to invest. Similarly, and as in many other industries such as the oil or minerals industry, "financial" markets are formed that are backed by this physical market, which allows certain producers or suppliers to hedge against a rise or fall in wholesale prices. This market, known as the futures market, also sees a volume of trades several times greater than the amount traded on the "physical" market, similar to what is observed on the oil markets.

2 Contribution to the public electricity service.

3 The taxes on the final consumption of electricity (in practice, CSPE and TCFE have merged under the name of TICFE: "*taxe intérieure sur la consommation finale d'électricité*"; in English, internal tax on the final consumption of electricity).

4 Fee contribution for delivery.

It is important to clearly distinguish two types of kWh costs, excluding access charges and excluding taxes: the average cost and the marginal cost. The average cost (levelized cost of electricity [LCOE]) is the cost price of a kWh or a MWh (at present value) obtained by dividing the sum of the fixed and variable costs incurred during the economic life of the plant by the total expected electricity production over the period. This presupposes that an assumption will be made about the power plant load factor during its economic lifespan (see Chapter 3 of Volume 1).

We then use this average cost as a basis to decide on an investment program: we choose the power plant with the lowest average cost. Thus, there are three variables that are decisive for the calculation:

– the amount of expenses over the lifespan of the power plant;

– the volume of kWh produced over the period;

– the chosen discount rate.

The marginal cost is constituted by the only variable cost of running the plant (excluding fixed costs); essentially, this covers the cost of fuel. On the basis of this marginal cost, the power plants are implemented to provide the electricity on the network (from within a given set of power plants). Once the set of power plants is known, we start by calling on the power plants that cost the least to operate to provide the electricity. As the demand for electricity increases, other power plants are used for which the marginal cost increases. This is what we call the merit order. On the wholesale electricity market (the so-called day-ahead market), the equilibrium price is aligned hour by hour with the marginal cost of the marginal power plant, which explains the volatility of this market when the marginal costs are very different from one power plant to another.

Run-of-river hydroelectricity stations, PV solar and wind power plants are called first (their marginal cost is almost zero). These are followed by coastal power plants, whose variable cost is low, then thermal power plants operating on coal, then those fueled by gas, and finally, "TAG" plants (gas turbines operating on gas or diesel) for peak periods. Hydroelectric dams also provide the needed electricity during peak periods since, in the case that electricity itself cannot be stored, water can be stored which can be used to drive turbines when necessary. The marginal cost of a hydroelectric dam facility is very low, but the value of the water used to produce electricity is very high at peak times when the risk of a blackout is approaching. This explains why these dam facilities are reserved for peak periods. This is also the case for pumping stations (stations for transferring energy through pumping [STEP]). In fact, the value of this hydroelectricity kWh is high, since it represents the avoided cost of the kWh not produced in the event of a blackout, in other words what is called the cost of failure (see Box 2.1). The marginal cost in this case incorporates the mathematical expectation of the avoided cost.

As demand increases (from off-peak hours of the day to full hours, then to peak hours), the marginal cost increases, and with it the equilibrium price on the wholesale market. Producers with marginal power plants therefore only recover the variable costs of their equipment; they recover the fixed costs of this equipment when their power plant is sub-marginal; in other words, when a power plant with a higher marginal cost becomes the marginal power plant of the market, allowing the balance between supply and demand. This provides them with a *mark-up*, an annuity that makes it possible to finance the equipment. The problem obviously arises for the marginal power plant during peak hours, which has difficulty covering its fixed costs. This is known as the *missing money* problem, highlighted by Cramton and Stoft (2006). If the equilibrium price were aligned with the only marginal cost of the TAG, this equipment would not be able to recover its fixed costs, which justifies a price higher than the marginal cost. This is the reason why the kWh must take into account both the variable cost and also the fixed cost of the TAG during peak periods. Gas turbines are reserved for the highest peaks since they can be brought online almost instantaneously, while their operating cost is high.

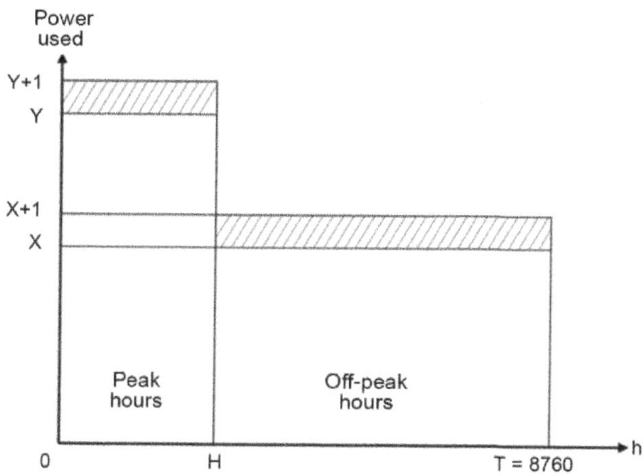

Figure 2.1. *Simplified representation of monotonous curve for applied power levels*

2.2. The price set by the merit order of marginal costs

To understand this logic of the merit order, it is necessary to start from the breakdown of hourly monotonous electric load. This is visualized schematically in

Figure 2.1 through the juxtaposition of two rectangles, where one represents the base power available (the period known as off-peak) and the other represents the additional power added at peak times (the peak period). In practice, the power used is a continuous curve. The reason the graph is discontinuous is for the sake of allowing simpler calculations (it is easier to determine the surface area of two rectangles than to take the integral of a continuous curve) (Hansen and Percebois 2019).

It is important not to confuse *differential rent* and *scarcity rent*. The former allows a producer, such as a nuclear producer, to recover its full costs if the rates are set on the basis of marginal costs; the latter grants it additional profits as soon as the prices allow it to receive revenues that go beyond its full costs. The logic of marginal cost pricing calls for the concept of differential or sub-marginal rent. *Scarcity rent*, as with *monopoly rent*, instead reflects a dysfunction within the competitive market.

Here, we will present a simplified grouping of production facilities, consisting of two types of plants exclusively: nuclear power plants for base production and gas turbines for peak production. The fixed cost per installed kW of nuclear power a is higher than the fixed cost of gas turbines (TAG) b, while the variable cost of nuclear power per hour of operation f is substantially lower than the variable cost of the TAG g, that is, $a > b$ and $f < g$.

The cost function of the nuclear kWh is given by $y = a + fh$, where h represents the number of operating hours. Such function for the TAG is given by $z = b + gh$. Thus, there is a special value of h written as H for which the two cost prices are equalized, or $y = z$ for $h = H$ with $H = (a - b)/(g - f)$ (the difference of fixed costs over the difference of variable costs).

Let us suppose that nuclear power production is in service all year round over the period (O, T) and that during the peak period (O, H), gas turbines are added. Here, X is the installed linear power (in kW) and $(Y - X)$ is the installed power in gas turbines (in kW). The total installed power is given by Y. Nuclear power is the marginal means of production at off-peak hours (H, T) and the TAG is the marginal equipment at peak times (O, H). The total cost of electricity production (the integral of the total consumption of power sources utilized) is given by:

$$C_0 = X(a + fT) + (Y - X)(b + gH) \qquad [2.1]$$

Now, we will examine the impact of an anticipated increase in electricity demand. We will suppose that there is an additional demand of 1 kW needed during either the peak, or the off-peak period, or the entire year (O, T). It is assumed that this demand is precisely anticipated by the producer, and that it adjusts its investment accordingly.

– *Case 1*: demand increases during the peak period (O, H). The total cost of production is now given by:

$$C_1 = X(a + fT) + \big((Y + 1) - X\big)(b + gH) \quad [2.2]$$

We deduce from this $\Delta C_1 = C_1 - C_0 = b + gH$. This is the marginal cost of the TAG that "sets the price" at peak times. All users consuming electricity during peak hours must therefore pay the kWh on the basis of the fixed cost and the variable cost of the gas turbine.

– *Case 2*: demand increases during the off-peak period (H, T). The total cost of production is now given by $C_2 = (X + 1)(a + fT) + (Y - (X + 1))(b + gH)$: The additional nuclear kW installed for off-peak hours to meet this additional demand is available at peak times and, if anticipation is possible, reduces the installed power in TAG by 1 kW. From this, we derive the marginal cost increased at off-peak hours, that is, $\Delta C_2 = C_2 - C_0 = f(T - H)$ if we replace H by its value $(a - b)/(g - f)$. At off-peak hours, the consumer must bear the only variable costs of the marginal facilities (in this case, nuclear plants).

– *Case 3*: demand grows all year round (O, T). In this case, it is necessary to invest in an additional 1 kW of nuclear power, and this power will be completely used. The installed power in TAG will remain unchanged, and therefore equal to $(Y - X)$. The total cost is given by $C_3 = (X + 1)(a + fT) + (Y - X)(b + gH)$, from which the additional cost borne by the producer $\Delta C_3 = C_3 - C_0 = a + fT$ is deducted. The result is therefore $\Delta C_3 = \Delta C_1 + \Delta C_2$.

The optimal pricing system, based on marginal costs, consists of recovering a revenue equal to $f(T - H)$ per kW used at off-peak hours and a revenue equal to $b + gH$ per kW called at peak times. Thus, we can see that the total revenue received by a nuclear reactor used all year round (O, T) is equal to $f(T - H) + b + gH$, that is, $a + fT$. By charging for nuclear power on the basis of the variable costs of nuclear power plants at off-peak hours and on the basis of the fixed costs and the variable costs of the gas turbine at peak hours, we cover both the fixed costs and the variable costs of nuclear power plants. The fact that the nuclear kWh is sold at a price that makes it possible to recover $b + gH$ per nuclear kW called on the peak period (O, H) does not therefore constitute an undue rent: it is the means of covering the fixed costs of nuclear power. If the price were to be set on the basis of the variable costs of the gas turbine alone at peak times, we would not recover all the fixed costs. An optimal price setting must follow the subsequent principles:

– At peak times, we must bear the variable costs and the fixed costs of the advanced equipment.

– At off-peak hours, only the variable costs of the marginal facilities used must be borne, the fixed costs of the base facilities are covered through the *differential rent* obtained at peak times.

– In practice, based on the available historical data, the peak period represents only the busiest 400 h of the year (out of 8,760). This translates to 18 hours a day for 22 days (if we exclude a 6 h period during the night), but without knowing beforehand when these 22 days will occur during the winter (between November and March). This mainly depends on the outside temperature, as the demand for electricity is particularly high during periods of extreme cold. It is considered that the peak period in a broader sense corresponds to the busiest 1,000 h of the year.

If for some reason (a producers' market power, speculation, excessive demand and under-capacity stock), the market price leads to a higher revenue than $b + gH$ at peak times, then additional revenue is generated, which can in principle translate either into a scarcity rent if the available electrical capacity is insufficient, or a monopoly rent if the price results from the use of market power (such as a withdrawal of capacity). This would be the case, for example, if some producers engage in capacity withholding, which would amount to voluntarily withdrawing available power plants from the network, for the sole purpose of raising wholesale prices and thereby generating a monopoly rent. This justifies greater vigilance on the part of the regulatory authorities, which must verify that a high market price does not result from such behavior. If fraud occurs, the penalties are severe.

Following a lack of capacity or a blackout of the network, the kWh not served have a cost much higher than the cost of the last delivered kWh. This cost depends on the depth of the failure η, which corresponds to the amount of energy not delivered by so-called "normal" means (in kWh) relative to the amount of kWh requested at a given moment. A failure occurs as soon as exceptional means are called on (such as an overload of thermal generators or a voltage drop). The cost of failure γ is a parabolic function given in the form of:

$$\gamma = a\,\eta^2 + b\,\eta + c,$$

where c represents the marginal cost of the last facilities used on the network just before the failure (at optimum levels if $\eta = 0$, we get $\gamma = c$). a and b are parameters determined by surveys at the time of a *blackout*. It is estimated that γ is in the order of €9,000/MWh, while the price of electricity on the wholesale market varies between €40 and €80/MWh depending on the period (based on figures observed until mid-2021, excluding exceptional situations such as those seen in 2022 with the rise in gas prices).

Box 2.1. *Concept of cost failure*

2.3. Nuclear power and the wholesale price of electricity

The goals of the European Commission in its "Energy and Climate Union" project were to accelerate the integration of wholesale electricity markets in Europe, continue the buildup of renewable energies and promote energy efficiency, all while placing European consumers at the heart of the energy transition. This transition seeks to gradually phase out carbon-based fossil fuels in favor of decarbonized energies, which requires a greater use of nuclear and renewable electricity. This means that we will see an increasing electrification of different activities, including in mobility. This transition is taking place today in the world in the name of preserving the environment and the fight against global warming.

In order to develop electricity exchanges in Europe and to promote the convergence of wholesale prices, the Union has adopted a set of rules and guidelines (known as network codes) concerning the allocation of interconnection capacities between countries and the management of congestion, making the solutions for connections between electricity markets legally binding. Market coupling consists of creating common exchange platforms and adopting mechanisms for implicit capacity auctions between countries, allowing better use of electrical interconnections and promoting the convergence of electricity prices. Initially set up between France, Belgium and the Netherlands in 2006, the coupling by wholesale market prices now covers 19 countries and constitutes an integrated exchange zone representing 85% of the continent's electricity consumption. The exit of the United Kingdom from the European Union, which took effect on January 1, 2021, resulted in the withdrawal of the country from the single energy market and the end of its coupling with European markets, acting as a return to explicit bidding on the use of interconnection capacities between the United Kingdom and the continent.

One of the consequences of the development of market coupling solutions, together with the increase in interconnection capacities, has been to reduce the price differences within the interconnected regions, even though significant differences may remain between regions. In its monitoring report published in February 2021, the Agency for the Cooperation of Energy Regulators (ACER) estimates that electricity price differentials within the Central-Western European markets (Austria, Belgium, France, Germany/Luxembourg and the Netherlands) were less than 1 euro per MWh for almost half of the time in 2019 (ACER 2020). This proportion has more than doubled in 5 years. The agency also observed that the differences between wholesale prices within the zone were less than 10 euros per MWh for about three quarters of the year. In some regions, such as the Baltic states or the Iberian Peninsula, electricity prices fluctuate for significant periods of time.

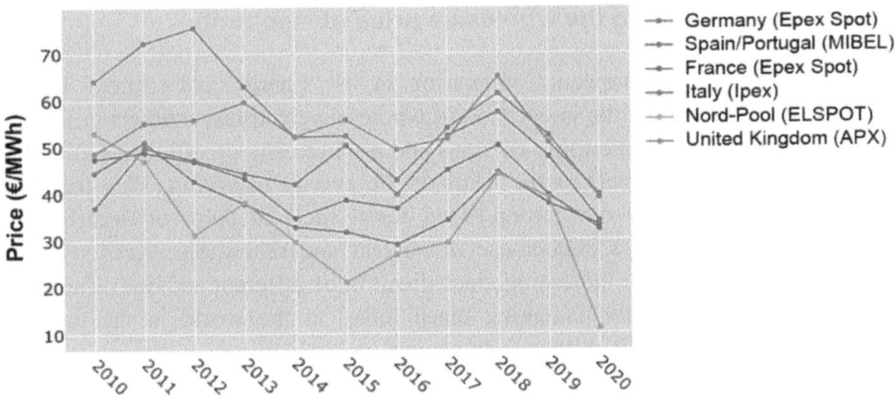

Figure 2.2. *Evolution of average base electricity prices on the European spot markets (source: Cyclops 2021). For a color version of this figure, see www.iste.co.uk/percebois/nuclear2.zip*

In Europe, electricity rates on wholesale markets have remained relatively stable on average over the past decade, or in some cases have even tended to decrease. If the beginnings of liberalization were marked by a sharp increase in electricity prices due to the rise in the price of fossil fuels between 2003 and 2008, the prices on the *spot* markets (Figure 2.2) have subsequently decreased due to, on the one hand, the stagnation of electricity consumption and, on the other hand, to the rise of renewable energies financed outside the market. This decrease in prices was particularly marked during the year 2020 due to the COVID-19 crisis and the restrictions on movement put in place by governments with the goal of curbing the epidemic. These measures led to a decrease in economic activity and electricity consumption, in the order of 15% in France during the first lockdown in March 2020 and a decrease of 3.5% over all of 2020, according to the calculations of the network manager (RTE 2021). In addition to the decline in electricity demand, the drop in oil and gas prices on the markets also contributed to the fall in electricity prices in Europe. Note that with the economic recovery, the increase in the price of gas and the price of carbon, wholesale electricity prices soared from the summer of 2021, exceeding 100 euros/MWh as of September 2021 (Figure 2.3). In 2022, wholesale prices exceeded 300 euros/MWh due to the surge in natural gas prices arising in particular due to the war in Ukraine.

The rates even climb to 620 euros per MWh on December 21, 2021. The very high volatility of the wholesale price arises when we compare the market situation on two dates 1 week apart, December 21 and 28, 2021. The 21st was a cold day, there was no wind in Europe (so the wind turbines did not produce much power), almost one-third of the nuclear power plants were unavailable (for reasons of scheduled maintenance or safety checks) and it was necessary to import a large amount of electricity to France. Thus, prices were very high (with a peak at 620 euros per MWh or 62 euro cents per kWh!).

It was the gas-fired power plants that were the marginal suppliers most of the time in Europe, so the price of electric MWh was set based on the very high price of gas (to which the price of coal was then added). France imported electricity the entire day, and the cost of that imported electricity was high, since it was very largely German thermal electricity produced using gas (or coal), which also had to take into account the price of CO_2. A week later on the 28th, the weather was mild and there was a lot of wind in Europe; economic activity was reduced during this period, the demand for heating was low, and consequently the demand for electricity had fallen sharply. Prices fell (the use of nuclear power even slightly so the wind turbines could be used more extensively) and France became an electricity exporter once again. Gas power plants were marginal for a small part of the time, and since the price of gas had fallen in Europe in the meantime, this caused the price of electric MWh to drop significantly (Table 2.1).

	December 21, 2021	December 28, 2021
Peak power	80 GW	62 GW
Share of nuclear power (daily average)	71%	65%
Share of gas power (daily average)	15%	5%
Share of wind power (daily average)	5%	18%
Gas prices in Europe	180 €/MWh	105 €/MWh
Spot price, Ministry of Electricity (France)	300 €/MWh	15 €/MWh
Maximum spot price of electricity (France)	620 €/MWh	150 €/MWh
Imports/exports from France	Importing 10 GW of thermal electricity at a high marginal cost which sets the price during the auction	Exporting 7 GW (mainly nuclear)
Observations	Gas is highly marginal in Europe	Gas is slightly marginal in Europe

Table 2.1. *Data relating to the production and consumption of electricity for December 21, 2021 and December 28, 2021 in France*

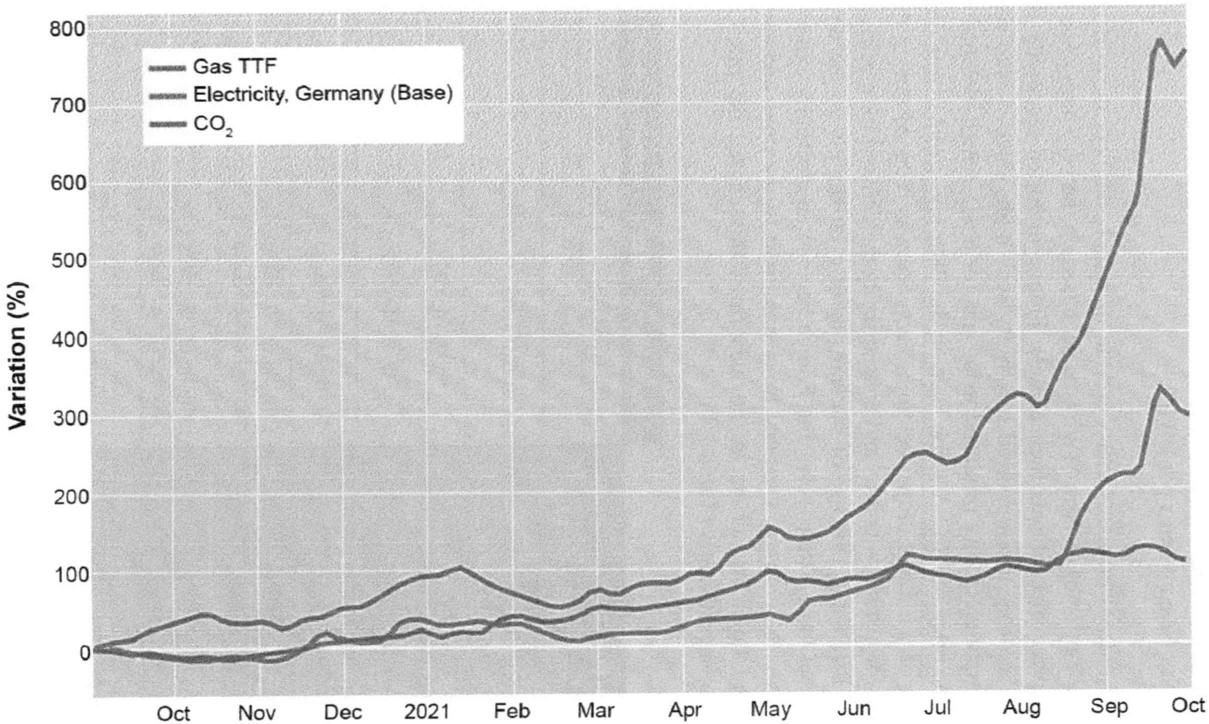

Figure 2.3. *Percentage change in the future prices of electricity, gas and CO_2 in Europe between September 2020 and September 2021 (source: www.tradingview.com). For a color version of this figure, see www.iste.co.uk/percebois/nuclear2.zip*

The level of economic activity, the weather conditions, the energy share and the price of gas are thus the four strategic variables that explain the price of electricity on the wholesale market during this period.

In order to achieve the objectives of the energy climate package, European states have put in place incentive policies seeking to promote the development of renewable energies. Different types of regulatory instruments, on the basis of either prices or quantities, have been used in Europe, including the system of guaranteed prices (*feed-in-tariffs*) with purchase obligations, which has by far constituted the main mode of support for renewables. In this type of regulation, a legislative body sets a guaranteed purchase price for green electricity, which is higher than the wholesale market price and accompanied by an obligation to purchase the kWh produced for a period of 15–20 years.

Under such a system, the producer's income does not vary; on the contrary, it is protected from fluctuations in the price of electricity on the wholesale market. This system has worked well, given that it has made it possible to achieve the objective of 20% renewables in the final energy consumption of the European Union, but it has proved costly in terms of public finances. In France, the extra cost of renewables has been passed on to all electricity consumers through a tax levied on energy bills (contribution to the public electricity service; *contribution au service public de l'électricité* [CSPE]), which was included in the general budget starting in 2016 and whose base has been expanded to fossil fuels. Public service charges related to the additional cost of electric renewables are expected to reach 5.6 billion euros in 2021 (CRE 2020). However, we note that France, whose share of renewable energies in final energy consumption was 17.2% in 2019 (SDES 2020), is behind the 2020 target of 23% (33% in 2030).

The production of renewable electricity (mainly via PV solar or wind), which benefits from a priority injection on the electricity networks, has tended to lower equilibrium prices on the wholesale markets in Europe in a context of stagnant demand. Since renewable energies are remunerated off-market, they are not penalized by the decrease in wholesale prices, which is not the case for other forms of production such as nuclear power or fossil power plants. Figure 2.4 illustrates the effect on wholesale markets of the supply of renewable electricity at zero marginal cost (wind and PV solar) which, when demand is unchanged, leads to a decrease in the volume of electricity sold by other electricity producers and a decrease in the market price (which remains set on the variable cost of marginal equipment). All producers remunerated on the basis of market prices are penalized, with the exception of renewable producers compensated through feed-in tariffs and who benefit from a priority of access to the network. This tends to accentuate the missing money problem on wholesale markets and to justify the use of capacity mechanisms to remunerate these producers who are called upon to serve as a back-up, due to the intermittent output of renewable energy production. Figure 2.4 shows how, following the injection of renewables at zero marginal cost, the translation of the merit order results in a fall in prices on the wholesale market (Hansen and Percebois 2019).

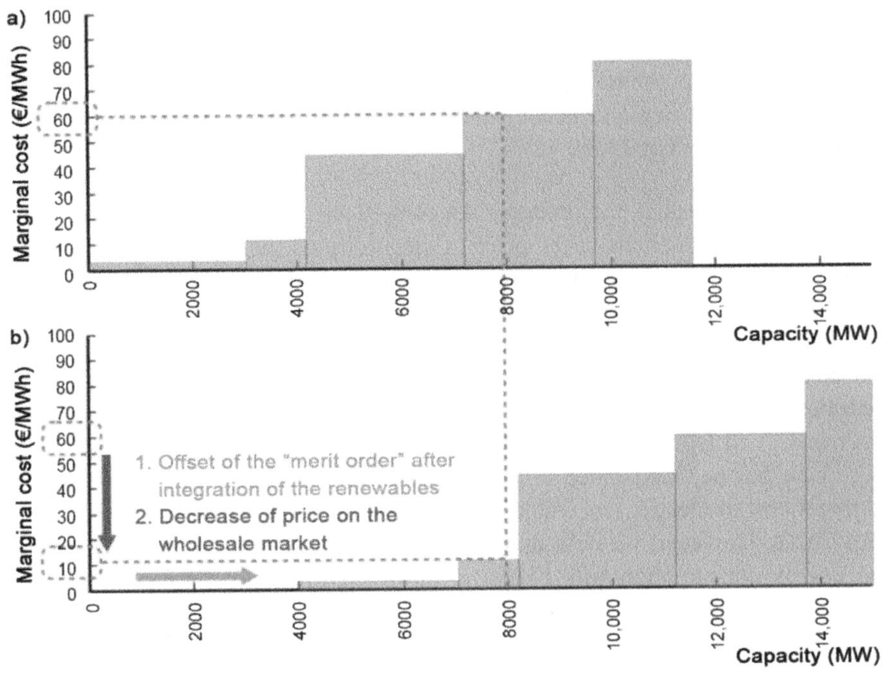

Figure 2.4. *Schematic representation of the translation of the merit order: (a) before injection of renewables at zero marginal cost; (b) After the injection of renewables, showing the fall in prices on the wholesale market. For a color version of this figure, see www.iste.co.uk/percebois/nuclear2.zip*

A paradox then arises with the development of renewables: as prices on wholesale markets decrease, the prices paid by the final consumer increase due to the charges related to the additional costs of renewables and the cost of back-up. The mechanisms involved are shown in Figure 2.5.

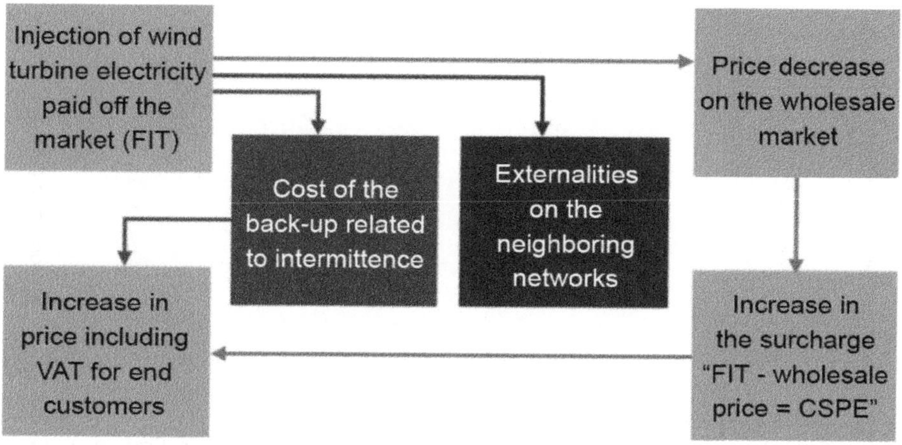

Figure 2.5. *Illustration of the mechanisms explaining the increase in the price including VAT for end customers after the addition of wind power paid for off the market. For a color version of this figure, see www.iste.co.uk/percebois/nuclear2.zip*

At certain times, the prices on the spot market can even become negative in the off-peak period when intermittent renewable production is strong. Renewable electricity producers, whose income is still mostly off-market, have no incentive to stop the supply of power to the grid, and in this case there is an oversupply of renewable electricity that must be discharged. The other producers on the spot market are meanwhile willing to pay operators in exchange for this service. For these producers, it is indeed preferable to endure a few hours at negative prices when demand is low rather than to stop production, given the start-up costs of the plants.

This phenomenon, which is relatively common in Germany when wind power production is high and demand is low, has also been observed in France since 2016. As shown in Figure 2.6, the occurrence of negative price periods in the French *spot* market has tended to increase in recent years under the effect of the integration of electrical markets and the increase in intermittent renewable production. The frequency of negative prices was particularly high in 2020 given the decrease in electricity demand caused by the lockdown measures.

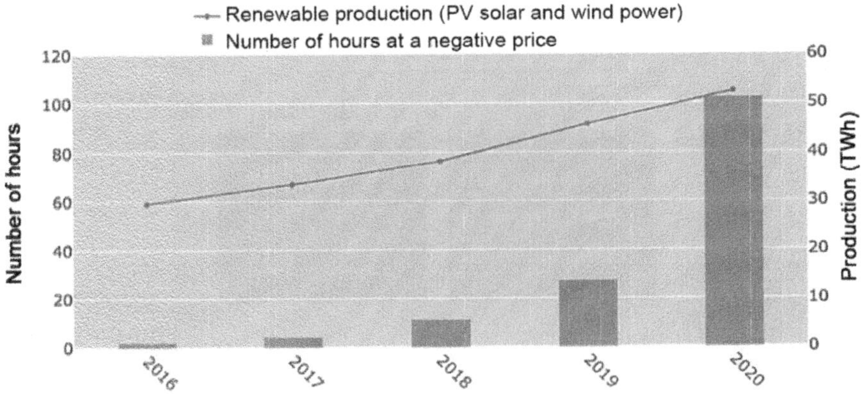

Figure 2.6. *Renewable electricity production and negative prices in France (source: authors based on data from ENTSO-E (European Network of Transmission System Operators for Electricity) and RTE). For a color version of this figure, see www.iste.co.uk/percebois/nuclear2.zip*

2.4. The need for a capacity market

The wholesale markets should in theory make it possible to ensure an identical depreciation and renewal of the production network through the sub-marginal rents paid by electricity producers. In practice, this has not always been the case in Europe. The price signal sent by the wholesale markets has often proven insufficient to allow the recovery of fixed production costs and guide long-term investment decisions. For this reason, some Member States have sought to set up parallel wholesale markets referred to as energy only, additional mechanisms aimed at compensating for capacities of production or removal in order to guarantee the security of supply.

Several systems have been proposed in Europe, and in particular we can distinguish price-based mechanisms, in which public authorities set the price of capacity and operators choose the level of investment, and value-based mechanisms where the market determines the level of compensation in order to achieve the target capacity. These mechanisms can either limit themselves to only certain capacities (new production capacities or specific technologies to be developed), or conversely, regenerate all the production capacities contributing to the security of supply. The price-based mechanisms, called "capacity payments", consist of directly providing remuneration to eligible electricity producers. This system, which has been experimented with by several countries (Spain, Greece, Ireland, Italy and Poland), is currently in the minority in Europe. Spain is the only country that still uses the capacity payment system, but only for power plants built before 2016.

Today, most countries make use of mechanisms based on their own resources, which can take the form of either targeted strategic reserves or markets open to all capacities. Strategic reserves consist of capacities that do not participate in the wholesale market, but are activated under certain conditions (in the case of shortages or price spikes on the wholesale market). This is the mechanism that has been established in Germany, Belgium, as well as the Nordic countries (Finland, Norway and Sweden). Other countries have opted for open capacity markets, where producers receive a specific payment for the available capacity in addition to the income derived from the sale of electricity on the wholesale market. The United Kingdom has established a centralized market of this kind in 2014, in which the network manager determines the capacity need to secure the supply several years in advance and proceeds to carry out auctions to acquire the corresponding capacities from electricity producers, or for back-up and storage operators. The winners receive the auction price in exchange for the capacity provided and must pay a penalty if they do not comply with their obligations.

France has chosen to set up a decentralized capacity market between operators of production or erasure capacities, on the one hand, and electricity suppliers (and network managers to cover losses), on the other hand. Concretely, the manager of the transport network (RTE) assigns capacity operators certificates called "capacity guarantees" (GC) attesting to the availability of their capacities during peak periods. These operators are then given the possibility to sell their certificates on the market to electricity suppliers, who are obliged to hold capacity guarantees covering the peak consumption of their customer portfolio during the winter period. As in the English system, operators have their capacities certified several years before delivery and the transfers of certificates are organized on the EPEX platform, which also manages the spot electricity market. The price of the capacity in both systems is quite close, at 20 €/kW/year on average for the French market in 2020 and around £10–15/kW/year for the English system. The prices of the capacity guarantee for deliveries in 2021 have increased sharply, reaching €40/kW/year on the French market and £45–50/kW/year in the English auctions.

Capacity markets thus offer an advantage for electricity producers, as they provide them with an income complementing that obtained on the wholesale market, and thus contributing to covering fixed costs. According to ACER estimates, the total revenue paid to capacity operators in Europe reached 3.9 billion euros in 2019 (ACER 2020). The most significant part, at almost 3 billion euros, is concentrated on the French (1.7 billion) and English (1.3 billion) markets, for a capacity cost estimated at more than €3/MWh, or about 9% of the average annual spot price of 2019. In total, these mechanisms offer a total of 185 GW of installed capacities, three quarters of which are made up of fossil fuel and nuclear thermal power plants. It should be noted that all means of production can participate in these capacity mechanisms, as long as they contribute to covering the peak of winter consumption, regardless of the controllable or non-controllable nature of production.

The problem is that, as far as the European Commission is concerned, the income granted to producers under the capacity mechanisms constitutes government assistance. That is why the Commission evaluates the measures proposed by the Member States in advance, most importantly by ensuring that these states do not introduce distortions of competition within the electricity markets. This means that the authorizations issued come with time limits. The Commission considers these mechanisms to be temporary devices that should make it possible to respond to an explicit supply security problem. In particular, the risk for the Commission is that capacity mechanisms are used to subsidize fossil fuel technologies, especially coal-fired power plants. To this end, it introduced a provision in the last legislative package adopted in 2019 (the Clean Energy Package) aimed at limiting the access of capacity markets to power plants emitting less than 550 gCO_2 per kWh. The Commission also advocates for these mechanisms, which are implemented at national level, to be open to capacity providers located in other Member States, as long as an interconnection exists, which in practice is rarely the case.

2.5. The carbon market and spot market in Europe

In parallel with the opening up of electricity markets to competition, in 2005 the European Union set up an emission allowance trading system (EU ETS, for European Union Emissions Trading Scheme) covering releases of CO_2 by power plants, refineries, heavy industries such as steel mills or cement plants and intra-EU air flights. The system works first by capping the total volume of emissions of the activities covered, then allocating the corresponding quotas to the participants concerned who can then exchange them on the market. The stronger the constraint, which is to say the lower the emission cap allowed in Europe, the higher the price of the quota on the market, and vice versa. Quotas are allocated to installations on an annual basis, which is done either free of charge, or as this becomes more and more the rule, at auctions. This is particularly the case for the electricity sector, which has had to buy all of its quotas at auction since 2013 (with the exception of a few Eastern European states authorized to distribute allowances free of charge to electricity producers beyond 2020). Each year, operators are obliged to return a volume of allowances corresponding to the verified emissions of their installations during the previous year. As they are the cause of more than half of the emissions covered by the scheme, electricity producers have been the first players in the market from the very beginning.

The carbon price in Europe has had the effect of increasing the production costs of fossil-based electricity, differentiated according to the carbon content of the MWh and penalizing the most emitting producers. At certain periods, this has led to more intense trade-offs between power plants on the wholesale market by encouraging operators to switch from coal-fired power plants to relatively lower emission gas-fired power plants, or to use biomass co-combustion in coal-fired power plants. Recall that to produce 1 kWh of electricity, a gas-fired power plant emits an average of 340–500 gCO_2, which is 50–65% less than a coal-fired power plant (IPCC 2006; IEA 2019). The price of the quota has also been able to encourage operators to optimize the management of their production networks, such as by making renovation investments to improve the thermal efficiency of power plants (De Perthuis and Solier 2019).

Generally more expensive to buy, but with less emissions, gas becomes more profitable than coal for electricity production once the price of the quota reaches a certain level. Once the level of the quota price is reached that makes it possible to activate the coal-to-gas switch at the level of the merit order, the electricity market is obviously highly dependent on the relative energy prices. An increase in gas prices relative to those of coal will have the effect of increasing the price for the coal-to-gas switch. This profitability gap also varies depending on the relative efficiency of converting gas and coal into electricity. Given that this parameter differs from one power plant to another in practice, it is customary to consider not one single price, but rather a gas-coal substitution price zone that takes into account the minimum and maximum values of the thermal efficiency of the plants. When the price of the quota on the carbon market is located within this zone, it means that there is at least one economically viable gas-coal substitution. Figure 2.7 shows the evolution of this substitution zone, with the price of the CO_2 superimposed above it.

As we can see, the price of the quota in Europe has often proved insufficient to trigger the switch from coal to gas. Placed at around €20/tCO_2 at the beginning of the second commitment period (2008–2012), the carbon price would fall rapidly following the crisis of 2009, which caused a drastic decrease in emissions and an excess of the supply of quotas on demand. This surplus of allowances would then increase due to the use of carbon credits generated under the compensation mechanisms of the Kyoto Protocol and policies to support the production of renewable electricity, which have contributed to reducing emissions and therefore the demand for allowances from the electricity sector. While the price of the coal-to-gas switch increased due to the rise in relative energy prices, the carbon price in Europe was permanently below the mark of €10/tCO_2. Between 2011 and 2016, it no longer allows coal to be taken out of the merit order, from then on playing only a limited role in the arbitration between the means of electricity production, much less than that of the relative price of energies and the guaranteed prices for renewables.

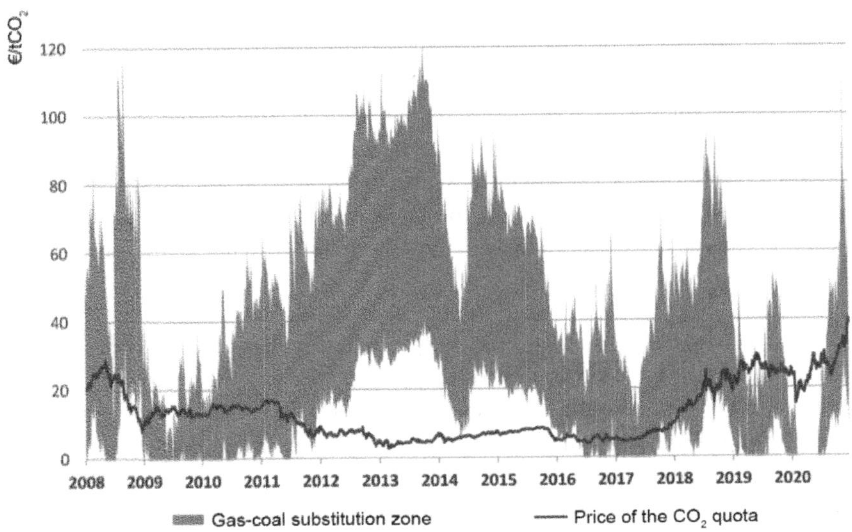

Figure 2.7. *Change in the coal/gas substitution zone in Europe from 2008 to 2020. The change in the price of the CO_2 emissions quotas is shown in blue (source: authors from ICE data). For a color version of this figure, see www.iste.co.uk/percebois/nuclear2.zip*

In order to correct these imbalances, the European Commission began a process of reform in 2012, which resulted in the establishment of a "stability reserve" (known as the Market Stability Reserve) in 2019, which worked to automatically modulate the volumes of allowances allocated to auctions according to the quantity of unused allowances in circulation. In concrete terms, this reform defines minimum and maximum thresholds within which the reserve intervenes by withdrawing quotas from auctions when the volume in circulation exceeds the maximum threshold and by re-injecting them into the market when the surplus of quotas falls below the minimum threshold. Its operation is also asymmetrical; the reserve absorbs quotas more quickly than it makes them available to market participants. In addition, in June 2021, the Member States agreed to raise the EU's 2030 emissions reduction target from –40% to –55%, which forms the next milestone on the road to carbon neutrality in 2050.

The "Fit for 55" package, presented in July 2021 by the European Commission, includes several measures that should allow for strengthening the quota market and extending the carbon price signal in Europe:

– strengthening the objective of reducing the emissions covered by the carbon market, which would go from a 43% to 61% emissions reduction in 2030 compared to 2005;

– gradually phasing in maritime transport emissions between 2023 and 2025 and ending the free allotments of quotas enjoyed by airlines for intra-EU flights;

– establishing a separate carbon market for emissions from the use of fossil fuels in the road transportation and building sectors by 2026;

– establish a carbon adjustment mechanism along the borders of the EU by 2026 in order to align the CO_2 costs of imported products with those supported by European producers. European importers of carbon-intensive products exposed to risks of leakage (cement, iron and steel, aluminum, fertilizers and electricity) will have to buy carbon certificates (the price of which will be indexed to that of the carbon market) in order to cover the emissions contained in the imported products.

Box 2.2. *"Fit for 55" Package and the carbon market reform*

To achieve this new objective, the European Commission has developed a series of proposals within the "Fit for 55" package presented in July 2021 (European Commission 2020), which seek in particular to strengthen the workings of the carbon market (see Box 2.2). By reducing the surplus of quotas and strengthening the medium-term reduction target, these reforms have made it possible to restore the confidence of participants and their expectations regarding the future constraints (Quemin and Trotignon 2021). As a result, the carbon price has gradually returned to market rates and at the beginning of 2021 has varied within a range of 20–40€/tCO_2, before crossing the €50/t mark in the summer of 2021. It was established at the end of 2021 at a level above 80 euros per ton, contributing to the surge in electricity prices on the wholesale market.

2.6. Accounting for the system costs in the merit order

The logic of the merit order on the basis of marginal costs leads to the creation of two types of distortions in the functioning of the wholesale electricity market, given that the so-called *external* costs or *system* costs are usually not taken into account. System externalities and costs are very close in this case.

2.6.1. *Crowd-out cost of controllable power plants*

This is primarily the crowd-out effect of the controllable power plants (nuclear power plants and power plants using fossil fuels) at non-zero marginal costs, replaced by non-controllable power plants (wind and PV solar) at zero marginal cost. The priority use of these intermittent non-controllable power plants leads to an increase in the average production cost of the controllable power plants since this reduces the load factor of these power plants. The problem is that non-controllable power plants are generally not available at peak times when market prices are compensatory, so they can only recover their fixed costs because they benefit from subsidies in the form of guaranteed purchase prices (feed-in tariffs) or additional remuneration in the form of bonuses.

The logic of bringing power plants into use in the order of their increasing marginal cost therefore generates distortions in competition if certain power plants are subsidized. If these power plants were not subsidized, they would not participate in auctions at zero marginal cost. Some consider that they should do this on the basis of their average cost. The replacement of controllable power plants by non-controllable power plants would then be economically justified if the average cost (and not the marginal cost) of producing the kWh by non-controllable power plants was lower than the marginal cost of the kWh produced by the replaced controllable power plants (Percebois and Pommeret 2018a; Dambrine 2019).

It should also be noted that fixing wholesale prices on the basis of marginal costs would lead to a price of zero or close to zero with a network consisting exclusively of renewable power plants at zero marginal cost. Based on market costs, the merit order could no longer ensure the financing of the plants. The compensation is then made exclusively through the capacity market.

The merit order gives different results depending on whether the average cost or the marginal cost of the kWh is used, as shown in Figure 2.8. If we take into account the average cost (LCOE), the entry into service of power plants ordered by increasing cost is as follows: hydroelectricity, nuclear, coal, gas, photovoltaic and then wind. If we take the marginal cost, the order of entry by increasing cost becomes hydroelectricity, photovoltaic, wind, nuclear, coal and then gas. This is logical since the marginal cost essentially accounts for the fuel costs. This cost is zero for hydroelectricity, wind and PV solar, and it is low for nuclear.

The mass-scale addition of wind and PV solar electricity causes the price to fall on the wholesale electricity market, which causes a loss of profit for all producers and can even generate negative prices during off-peak periods. It is the fatal nature of this electricity that is the main cause of these negative prices. It is necessary to pay operators so that they agree to acquire electricity that has become cumbersome. The logic is that interrupting the supply of this electricity if the demand is low and if this risk leads to negative market prices. But renewable energy producers, who are paid according to the volumes supplied, reject this solution.

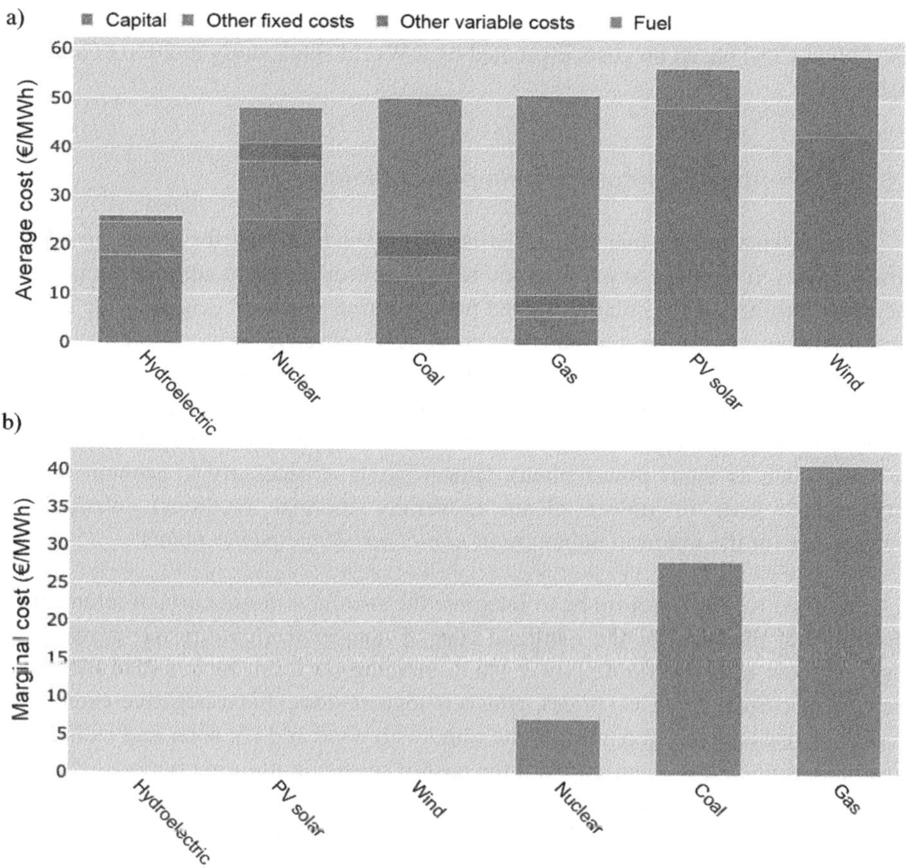

Figure 2.8. *Merit order obtained by the average cost method (a) and the cost of the fuel alone (b) (source: Percebois and Pommeret 2021a). For a color version of this figure, see www.iste.co.uk/percebois/nuclear2.zip*

In other words, the fixed cost per *non-controllable* kWh must be lower than the variable cost (fuel cost and carbon price) of the *controllable* kWh if we wish to eliminate distortions in competition. The replacement of *controllable* power plants, apart from this specific case of failed costs (or stranded costs), means that the nuclear and fossil-powered thermal operators cannot recover all their fixed costs. The study by Percebois and Pommeret (2018) estimated the nuclear shortfall linked to this replacement at several billion euros per year.

We note, as shown in Figure 2.8, that the average cost of wind and photovoltaic power is higher here than the marginal cost of the kWh produced with coal and gas (excluding CO_2) based on costs estimated by a World Bank study in 2020 (Timilsina 2020).

2.6.2. *Back-up of non-controllable power plants*

Another *system* cost that is ignored is that of the back-ups for *non-controllable* power plants. It is necessary to provide reserve power plants to address the issue of the unavailability of PV solar and wind power during what are generally the busiest periods of the year. One alternative is to store electricity, so as to later release it when necessary. This can be done through short-term batteries and through *methanation* over the short-to-medium term (using the electrolysis of water to produce hydrogen, which is then combined with CO_2 to obtain methane, then used in plants such as a gas power plant). Otherwise, it is necessary to account for the maintenance cost of power plants (generally thermal facilities) intended to compensate for the intermittent nature of *non-controllable* power plants.

The logic would therefore be to integrate the *storage/release* costs of intermittent electricity at the level of the marginal cost of *non-controllable* power plants. The marginal cost of *controllable* power plants running on fossil fuels already includes the cost of carbon (the CO_2 price), which is logical since it is a negative externality that gives rise to costs. We must do the same with wind and PV solar and include the cost of storing excess electricity and the cost of releasing it during periods when PV solar and wind are not available. The cost of nuclear waste management, proportional to the volume of fuel that is treated, must therefore also be included in the marginal cost of nuclear kWh, even if it is found to be low in the data presented in Court of Auditors (2012) in the order of €7/MWh.

The study by Percebois and Pommeret (2019), updated in 2020 (Percebois and Pommeret 2021b), with recent data from the World Bank (Timilsina 2020) shows that the merit order based on average costs (LCOE) gives the following results: run-of-river hydroelectricity power plants are given top priority, followed by nuclear power plants, coal-fired power plants and then gas-fired power plants. Photovoltaic and wind power plants are used in the last step (see Table 2.2 and Figure 2.9).

The updated average cost of MWh (LCOE) is the sum of four elements (see Table 2.2 for details):

– the annualized cost of capital (ACC);

– the fixed costs of maintaining operational conditions (FCMOC);

– the variable costs of maintaining operational conditions (VCMOC);

– the fuel costs (FC).

	The cost of capital (€/MWh)	Other fixed costs (€/MWh)	Other variable costs (€/MWh)	Fuel costs (€/MWh)	LCOE (€/MWh)
Photovoltaic	48.49	7.93	–	–	56.42
Wind power	42.69	16.51	–	–	59.20
Gas	5.81	1.72	2.67	40.72	50.93
Hydroelectric	18.13	6.70	1.14	–	25.98
Coal	13.24	4.59	4.46	27.80	50.09
Nuclear	25.33	11.80	3.78	7.24	48.15

Table 2.2. *Breakdown of the LCOE for different types of electrical production (source: Percebois and Pommeret 2021b)*

If the merit order is based on marginal costs excluding externalities (fuel cost and proportional variable cost), the order changes: it becomes hydroelectric power plants, photovoltaic power plants, then wind power plants that are prioritized, followed by nuclear, coal and gas. If we introduce externalities (*system* costs), the order is not changed, as long as the storage/release costs remains low (€15/MWh), but the equilibrium market price will tend to rise to integrate the cost of these externalities.

As soon as this storage cost exceeds €20/MWh (a cost currently estimated to be at least 30–40 euros in the Alps), nuclear power is used before renewables. Similarly, gas-fired power plants are used before coal-fired power plants, and no longer after them, as soon as the price per ton of CO_2 exceeds 20–25 euros. It should be remembered that the carbon content of a MWh produced with coal is more than twice as high as that observed in a combined cycle gas power plant.

Figure 2.8(a) shows the merit order according to the increasing average cost (LCOE) of the centers used. With the assumptions that are adopted, the LCOE of renewables is higher than that of fossil fuels and, *a fortiori*, than that of nuclear, but the difference is not very large. On the other hand, the structure of average costs is very different from one energy source to another.

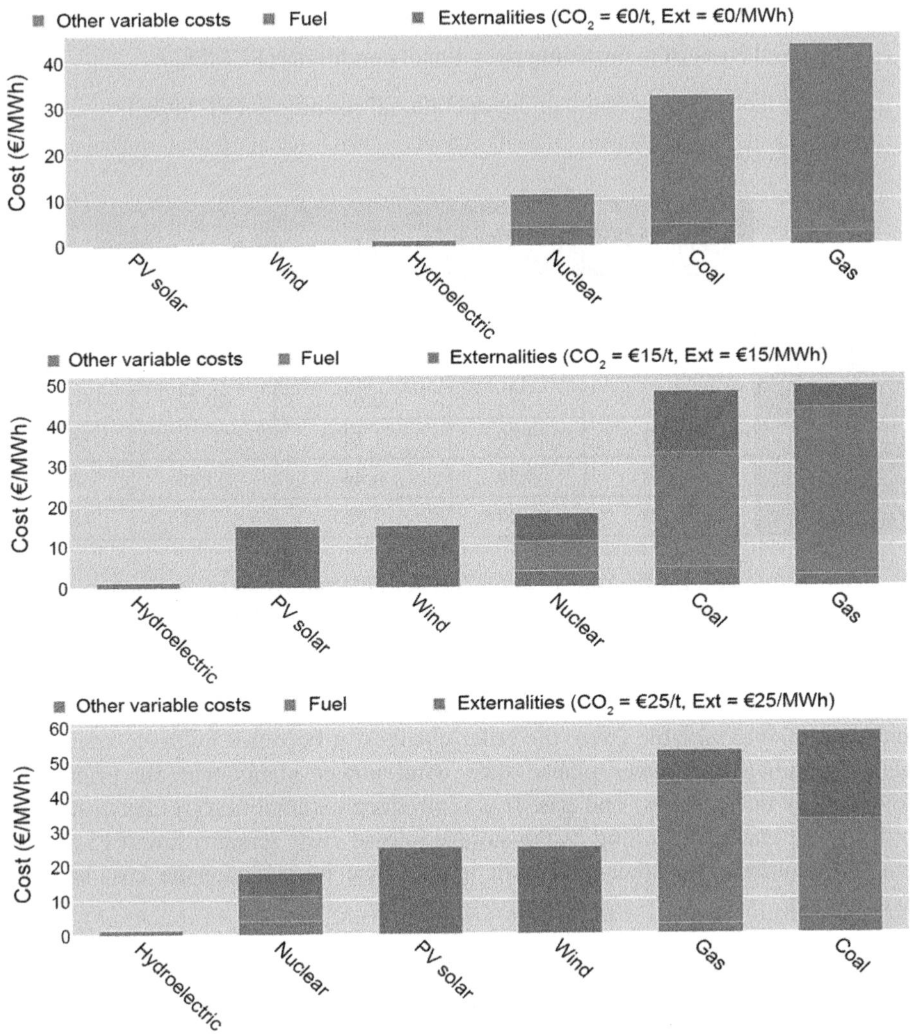

Figure 2.9. *Merit order based on assumptions about the price of CO_2 and the cost of storage (source: Percebois and Pommeret 2021b). For a color version of this figure, see www.iste.co.uk/percebois/nuclear2.zip*

The graphs in Figure 2.9 give the merit order on the basis of marginal costs by first describing this merit order based on the marginal cost of power plants, when only fuel costs and proportional variable costs are taken into account. The following graph gives this same merit order when we retain the full marginal cost with the *system* costs, that is to say, when we count the externalities based on one ton of CO_2 at 15 euros and the cost of storing the MWh also at 15 euros. The graph above describes the merit order when the price per ton of CO_2 is 25 euros, while the storage cost of the MWh is also increased to 25 euros.

The use of power plants on the basis of marginal costs is different from the order of merit of average costs: renewable power plants are prioritized as long as the cost of storage is not too high and, *a fortiori*, if the carbon price is high. With a low carbon price and a high cost of storage, it is to be expected that the order will be reversed. This implies that there is a *pivot value* for the price of carbon that guarantees that renewables will be implemented before fossil fuels, and this value varies with the cost of storing intermittent energies. The higher the cost of storage, the higher the carbon price must also be for renewables to maintain this priority.

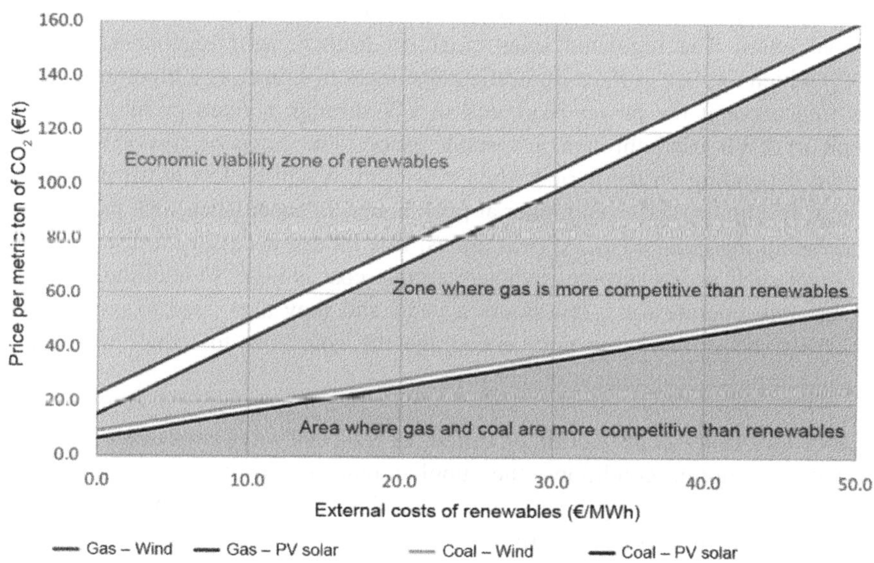

Figure 2.10. *Economic competitivity zone for renewable energies, gas and coal, according to the price per ton of CO_2 and externalities of renewable energies (source: Percebois and Pommeret 2021b). For a color version of this figure, see www.iste.co.uk/percebois/nuclear2.zip*

Thus, the consideration of externalities is not neutral with regard to the merit order based on marginal costs. Obviously, this consideration translates into an increase in the equilibrium price of MWh on the wholesale market. Consumers must pay for the externalities arising from their patterns of consumption. We note that in return, certain taxes intended to finance renewables, such as the CSPE, could be reduced.

From this, we can deduce the *pivot values* that cause the choices to switch over. Figure 2.10 shows an area where electricity produced with gas and coal is more competitive than the electricity produced with intermittent renewables, then an area where electricity produced using gas maintains its competitiveness (in this case, coal has been eliminated), and finally, an area where renewable electricity is more competitive than its competitors, either because the carbon price is high, or because the storage/release costs are low.

2.6.3. *Rates based on costs*

The price paid by end consumers will take into account this variability of marginal costs. The regulated sales tariff (in French: *tarif réglementé de vente* [TRV]) which applies to most domestic consumers in France is a binomial rate that takes into account the power requested in kW through a fixed premium, and the amount of kWh used through a variable price. The price of this kWh will be different depending on the period when it is used: it will be low at off-peak hours, higher at full hours, and even higher at peak hours, in accordance with the principle of increasing marginal costs. At peak times, it is indeed necessary to use expensive equipment such as gas turbines, whose variable cost is high. In addition, these gas turbines only operate for a few hours a year, and they also need to recover their fixed costs. Note that some rates encourage the consumer to reduce use at peak times in return for discounts at other times.

Since electricity is not currently stored, at least not on a large scale and under reasonable economic conditions, the supplier must maintain a network of power plants capable of meeting peak demand, which is expensive in overall terms. Otherwise, it is necessary to offload part of the demand, which is even more expensive for consumers, or worse still, a potential blackout, which would be very detrimental to the country's economy. Demand management through digital techniques (such as the Ecowatt system) is used increasingly to maintain the balance between electricity supply and demand by erasing this demand during peak periods.

2.7. The case of nuclear power and the ARENH mechanism, a hindrance to the market

2.7.1. *The ARENH, a "canal lock" for removing rents*

2.7.1.1. *The principle of the ARENH*

Devised by the Champsaur Commission in 2009 (Champsaur 2009), the ARENH (*accès régulé à l'énergie nucléaire historique*, "regulated access to historical nuclear energy") is a mechanism implemented by the NOME law, which passed in 2010 and entered into effect in July 2011. The purpose of this mechanism was to allow EDF's competitors, the "entrants", to be able to compete on equal terms with the incumbent in a competitive field which, since July 2007, allowed all European electricity consumers to choose their supplier. The European Commission, for the purpose of opening up to greater competition, then imposed on the historic operator EDF, which benefitted from a dominant position on the French market, a loss in market share to its competitors due to the competitiveness of nuclear kWh.

With the rise in oil prices, and consequently natural gas prices in 2008, electricity prices on the wholesale electricity market had risen sharply as thermal power plants running on gas or coal, fuels whose prices were influenced by the price of oil, were price makers in many cases. Since the marginal cost of a thermal power plant is much higher than that of a nuclear power plant, this gave EDF a comparative advantage, a "rent", and prevented the "contractors" from gaining market share. This rent has the nature of a scarcity rent, linked to investments made before the opening to competition. The average cost price of the MWh produced by a largely amortized nuclear power plant (which is the case for the 56 reactors currently in operation in France since the two Fessenheim reactors were shut down in 2020) still remains competitive if we compare it to the average cost of thermal electricity, as we have seen above.

Two solutions were presented to the Champsaur Commission: either penalize EDF by taxing the company with an amount representing the so-called nuclear scarcity "rent" and which would in practice provide an advantage linked to past public choices, or help the "entrants" to acquire nuclear electricity on the basis of the real cost from which EDF benefited, a cost lower than the wholesale market price. This was a "canal lock" system, which in first case amounted to raising EDF's costs to match those borne by its competitors (the wholesale market price), and in the second case to bringing down those of its competitors to the level of the advantageous cost borne by EDF. The Champsaur Commission recommended opting for the second solution, but limiting this "access to historical nuclear power" to a "reasonable" level depending on the portfolio of customers available to the "entrants": 100 TWh of the 400 TWh of nuclear power (or about 25% of the annual nuclear production at the time) and continuing until 2025. Following the recommendations of the second Champsaur Commission, the price was set by the public authorities at 40 euros per MWh for the last 6 months of 2011 and at 42 euros starting on January 1, 2012, a rate that was still in effect in 2022.

The "entrants" have the option (which is not an obligation) to purchase certain amounts from ARENH in MWh, at regular intervals, pro-rated from the consumption of their customers within the national territory. They were required not to resell these volumes on the wholesale market and to allow their customers to benefit from the advantageous price obtained.

The existence of the ARENH reduced the volume of electricity exchanged on the wholesale market by the same amount, since it is an off-market exchange at a regulated price. This has the potential to increase the volatility of prices on this market. In practice, given the importance of nuclear power in electricity production in France (accounting for 69%–70% of it), nuclear power is partly sold at market rates and partly at regulated rates.

2.7.1.2. *Setting of the ARENH price*

The structure of the ARENH price is illustrated in Figure 2.11.

According to the Champsaur Commission II (in its non-public report), the ARENH award must reflect the production costs of the historical nuclear power plant, distinguishing between operational expenses, future investments and past investments.

Operational expenses must be passed on "over time" (following the *pass-through* logic). These have several elements. First of all, they include the costs related to the nuclear fuel consumed during the year (including the calculation of the downstream costs of the cycle). Next are the purchases committed to be made for the operation and maintenance of the 58 nuclear units (56 in 2022). Other costs include personnel costs, then taxes and levies (excluding corporate taxes) borne by nuclear power plants. These expenses represent about 10 billion euros/year, or when reduced to a production of 420 TWh, about € 25_{2011}/MWh.

Future investments concern the maintenance and possible extension of the current network (the reactors will turn 40 years old between 2020 and 2030). At the time, these investments were estimated at between 3 and 4 billion euros in 2011 per year, but they are expected to increase. We can take into account either the repayment annuities, with the invested capital being paid at a rate equal to the

WACC (Weighted Average Cost of Capital) before tax (following the logic of the "economic lifespan"), or consider these investment expenses in the same way as the operational expenses and pass them to the price in the year in which they are carried out or the following year (following the logic of "liquidity"; the investments will be financed regardless of the operating period ultimately authorized by the ASN, but EDF may be tempted to anticipate certain investments at the end of the period, i.e. 2025). In the event that the investments are passed through as operational costs (the solution chosen), they would be in the order of 5–8 €$_{2011}$/MWh at the beginning of the period and would reach €10$_{2011}$/MWh at the height of the implementation of the program to extend the service life of the 58 reactors (56 in 2022). Finally, the commission settled on €8/MWh.

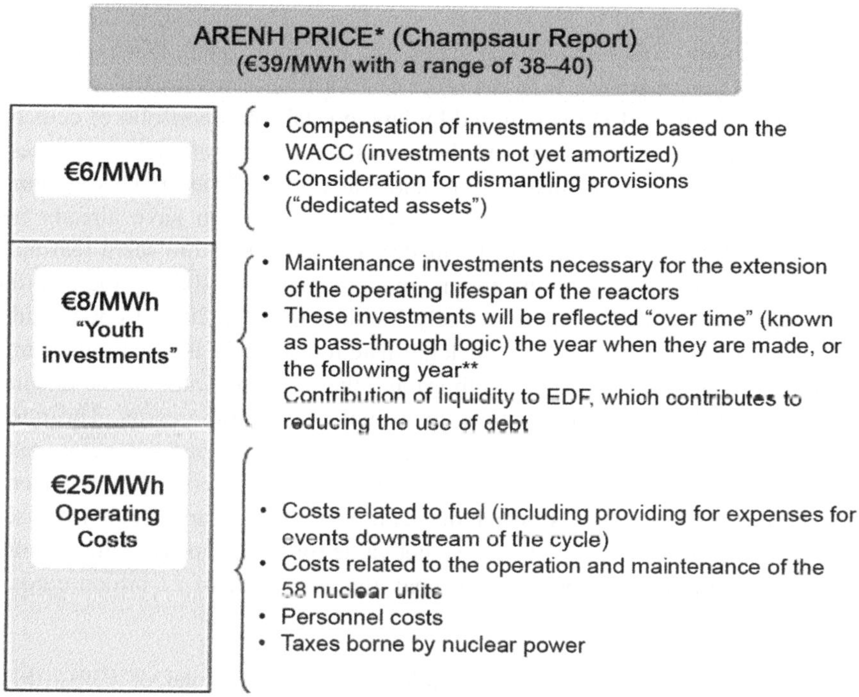

*Before taking into account any additional costs related to the Fukushima accident (security-related costs).

**The repayment and the remuneration of the capital amount could be spread over the lifetime of the assets (which is unknown from the outset) with the repayment rents accounted for in the ARENH. The two methods give slightly different results in practice. Over time, the investments are paid for regardless of the duration of operation authorized by the ASN.

Figure 2.11. *ARENH price structure (source: Champsaur 2009, 2011). For a color version of this figure, see www.iste.co.uk/percebois/nuclear2.zip*

Past investments must be compensated and reimbursed to allow EDF to proceed with the eventual renovation of the network (keeping in mind that the ARENH cannot provide the financing of future plants). The commission proposed to build an asset base covering the residual capital still left immobilized in the historic nuclear plant network (at the time, this network was 26 years old on average, midway through its lifespan). Earlier investments have already been partially repaid through the sale of electricity since the mid-1990s. The issue then becomes that of determining the residual share of the remaining capital to be remunerated and repaid (debt and equity).

The capital already repaid represents the amounts amortized. At the beginning of the lifespan of the plants, the depreciation, calculated according to a decreasing method and over only 30 years, was very significant. The interest on the debt was paid and the equity was remunerated. Thus, the net book value, that is to say the investment not yet amortized, is in the order of 15 billion euros in 2011 (capital not yet repaid). In addition, EDF is required by law to establish a portfolio of dedicated assets for an amount corresponding to the provisions for the long-term expenses of the historical plant network (dismantling and waste management). This amount is estimated at 18 billion euros in 2011 (of which 15.3 billion have already been acquired, and 2.7 billion remains to be made). It would therefore seem reasonable that the price of the ARENH allows for an amount equal to 15/40ths of the 18 billion euros to be provided, that is, approximately 7 billion euros in 2011, to be repaid by 2025 (over the next 15 years). It seems legitimate to use as an asset base the sum of the share of dedicated assets (i.e. 7 billion) and the net book value of the historical network (i.e. 15 billion), which comes to a total of 22 billion euros. The rate of remuneration for the capital retain was at that time a real rate of 8.4% before tax. It is assumed that all past costs are repaid at the end of the period (2025), over an amortization period of 15 years, which makes it possible to clear the debt (in case of extension of the network, this will leave a net value for the company). The economic annuity that would make it possible to repay the asset base of 22 billion euros by 2025 is in the order of €6/MWh.

The "fair" price of the ARENH should therefore be a total of €39_{2011}/MWh (25 + 8 + 6). The price is certainly about €5 to €7/MWh higher than the cost of the underlying historical nuclear power in the regulated rates of the time (TRV), but the difference is justified by the investment effort to be made in the coming years to extend the life of the fleet (at the time the report was delivered in March 2011, the Fukushima accident had occurred, so there would be additional security investments that would have to be committed at the request of the ASN). The price of the ARENH, which would be consistent with the TRV in force at the time, should have been about 32–34 €/MWh. But due to the investments to be planned, the price of the

ARENH had to reach €39/MWh (or even more) starting in 2015. A starting price of €39/MWh in 2011 would thus correspond to an advance of the level of costs to be covered in 2015.

The price of the ARENH was consistent with the TaRTAM (the provisional regulated rate based on the old TRVs with an additional discount) was therefore estimated by the commission at €39/MWh. It is necessary for an alternative supplier to be able to offer a former TaRTAM consumer a comparable price offer while covering its costs. This consistency with the TaRTAM obviously depends on the allocation rate of the ARENH volume and the wholesale market price, which are two exogenous variables (at the time, the wholesale price of MWh was in the order of base rates of €60 and around €76 for peak rates on average). A difficulty arises from the fact that there were several TaRTAM values (with high divergence around the average; the commission considered an average portfolio).

The price of the ARENH, consistent with the TaRTAM, means that for suppliers wishing to compete with the TRV (especially small consumers), supply costs were higher than the selling price (the so-called *rate chisel* effect). Thus, the idea that was proposed to compensate for this was by increasing the volume of ARENH allocated to suppliers for small consumers. A *rate chisel* happens when a company in a dominant position sets wholesale prices higher than its retail prices, thus preventing competition from forming. The "TaRTAM charge" borne by EDF to compensate suppliers disappears with the entry into force of the ARENH. Until then, suppliers delivering customers to the TaRTAM were compensated for the additional cost they incurred, up to the difference between the higher wholesale market price and the TaRTAM. This compensation was financed by a tax on nuclear power and hydroelectric power (paid almost entirely by EDF). Thus, on the basis of an ARENH starting price of €39 and the definition of the compensation mechanism, EDF received a de facto ARENH price ranging from 41 to 42 €/MWh. We note that in 2015, ARENH prices and regulated rates were required to be merged (by adding the various costs). After Fukushima, the ministry decided to set the rate of €40/MWh for the last 6 months of 2011, and €42 starting in 2012. The Ministry considered that the new security expenses related to Fukushima justified raising the price proposed by the Champsaur Commission somewhat.

2.7.1.3. *The new data*

What neither the Champsaur Commission nor the public authorities imagined was that the wholesale market price would collapse a few years later, and sometimes remain permanently lower than the level of the ARENH price. The demand for ARENH therefore tended to fall, especially in 2015, even reaching zero in 2016. There were several reasons for this new situation. Starting in 2014–2015, oil prices fell, along with gas prices But above all, the massive subsidies granted throughout Europe to renewable energies, such as PV solar and wind, caused prices to fall on the wholesale electricity market. This renewable electricity, financed outside the market through guaranteed prices (feed-in tariffs), participates in auctions at its marginal cost at close to zero, which, due to high overproduction of electricity in a context where demand remained stable, led to very low and sometimes negative wholesale prices. The public authorities even had to introduce a "capacity market", as we examined previously, to allow operators to finance the fixed costs of their power plants and thus ensure that the power production called for would be sufficient to pass the winter peak and avoid blackouts. As a result, the demand for ARENH declined, even reaching zero at times, when the wholesale market prices were lower than the regulated ARENH price. This occurred in 2015 and 2016, as shown in Figure 2.12. It then rose again with the rise in wholesale prices.

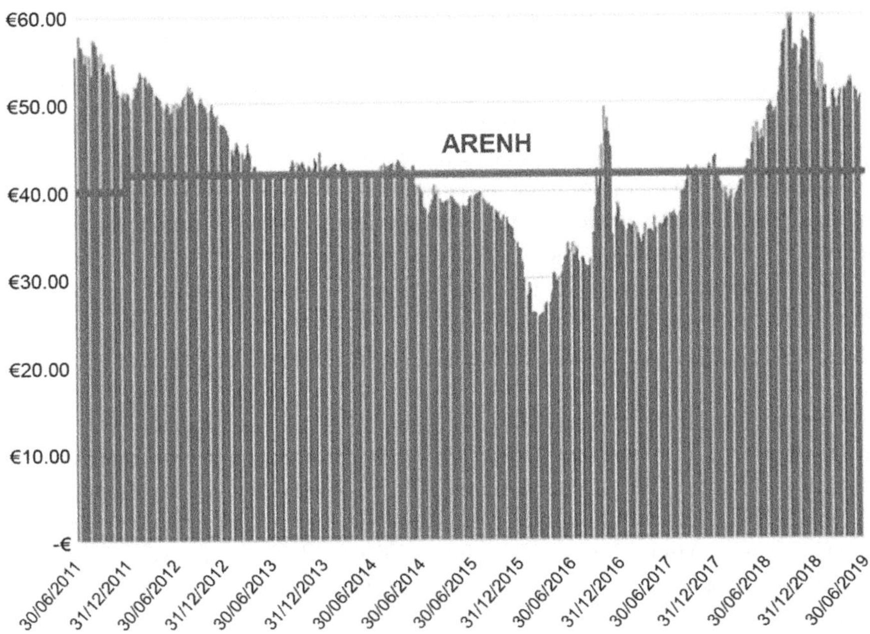

Figure 2.12. *Change in the wholesale electricity price since the establishment of the ARENH (source: Epex Spot data). For a color version of this figure, see www.iste.co.uk/percebois/nuclear2.zip*

The ARENH demand for 2019 surpassed the threshold of 100 TWh (133 TWh) for the first time, and it was necessary to reduce the claims of each of the alternatives by using the *pro rata* technique. Recall that ARENH requests must be justified by the volume of electricity purchased by customers in France. This was also the case for 2020 when subscriptions were made at the end of 2019 (146 TWh). At that time, the entrants anticipated an increase in the wholesale price of electricity due to the expected increase in the price of oil, the sharp increase in the price of the CO_2 quota which had already been observed – increasing from 5 to 25 euros in a few years – and the accelerated closure of coal-fired power plants in Europe (particularly in Germany). As a result, they sought the ability to acquire 150 TWh, but EDF was not in favor of this unless it received a significantly higher compensation in return. The pandemic that broke out in early 2020 shook up the situation and caused a dispute between EDF and some of its competitors.

The COVID-19 crisis caused the price of electricity on the wholesale market to collapse due to the fall in electricity demand resulting from the drop in economic activity during the lockdown (Percebois and Solier 2020a). Some alternatives that at the end of 2019 had anticipated a price increase on the wholesale market invoked the clause of *force majeure* due to not respecting the contract signed at that time. The CRE contested this request, and the Council of State ruled in favor of the CRE. But on the basis of specific provisions of Article 10 of the framework contract, which is binding between EDF and its alternatives, the commercial court of Paris ruled in favor of the alternatives, which had appeared before it in summary proceedings on March 27 (in particular TDE, Total-Direct Energy, and the AFIEG). EDF appealed, but the Court of Appeal upheld the commercial court's ruling on July 28, 2020. Due to expectations that wholesale prices would increase, the demand for ARENH for 2021 reached 147 TWh, which once again forced the CRE to use the *pro rata* method to stay within the 100 TWh range.

2.7.1.4. *Force majeure*

Electricity prices on the wholesale market fluctuated on average between 10 and 30 euros per MWh between mid-March and July 2020 due to the sudden drop in electricity demand as a result of the pandemic. These prices were even sometimes negative, as was the case for the price negotiated on Sunday, May 31 at 3 p.m. (–40.8 €/MWh). They rose slightly in the second half of 2020, and then spiked at the end of 2021 after the strong global economic recovery and the rise in gas prices that resulted (gas facilities were often in a marginal position on the wholesale electricity market).

The order of the commercial court of Paris was made public on May 20, requiring EDF to suspend the framework agreement that bound the company to TDE. However, the disruption of electricity deliveries under the ARENH only occurred up to the level of the amount by which the consumption of the portfolio of end consumers held by TDE decreased. A large proportion of TDE's customers had not purchased the negotiated volume of electricity due to the lockdown declared by the public authorities.

Since electricity is not stored, TDE had to resell the volumes of ARENH electricity that were negotiated and not acquired by its customers at a loss on the wholesale market. A very simple model will allow us to understand the scope of the decision (Percebois and Solier 2020b).

Let Q_1 be the volume of ARENH negotiated by TDE, Q_2 the volume of ARENH actually sold by TDE over the period, p_a the price of the ARENH and p_1 the spot price of electricity. TDE paid Q_1p_a to EDF, but only sold Q_2p_1. The logic of this "free option" provided by the ARENH is that TDE supports the differential $Q_2(p_a - p_1)$: this is the "market" risk of any contract of this type. What the court is asking is that TDE be reimbursed not for the amount $(Q_1 - Q_2)\ p_a$ but the amount $(Q_1 - Q_2)\ (p_a - p_1)$, where this electricity that is not sold to customers has been sold at the wholesale market price, much lower than the ARENH price. In other words, TDE must bear the "price" risk for the amounts of electricity sold, but not the "volume" risk linked to the fall in the consumption of its customers following the health crisis. The court considers that TDE was required to deliver the quantities of ARENH that were ordered, that is, Q_1, and was forced to sell at a loss on the wholesale market what the company could not sell to its customers, that is, the electrical volume of $(Q_1 - Q_2)$. The unit loss of TDE is equal to $(p_a - p_1)$.

Note that the "price" risk $Q_2(p_a - p_1)$ is perhaps hypothetical and has not been proven, insofar as many suppliers sell their electricity on the basis of a futures price or a fixed price that may be higher than the spot market price. This raises the question of the relevance of the wholesale price as a criterion for evaluating the gains and losses of both parties. The loss suffered by TDE on the volumes sold Q_2 may be less than this amount if the unit price that is actually negotiated by TDE is greater than p_1.

The argument of the commercial court, which rules on law more than equity, is mainly based on a very specific provision of the framework contract, that of Article 10, which provides that *force majeure* applies *as of right* given the occurrence of an "impossibility to perform the obligations under reasonable economic conditions". We recall that *force majeure* generally requires three conditions: unpredictability, independence, and inevitability. Nobody disputes the independence of the events (TDE is not responsible for the decision of the confinement which is a decision of the State) or their inevitability (TDE could not force its customers to consume its electricity, and does not have the means to store the electricity that is not purchased; it was therefore necessary to sell it at a loss). On the other hand, the unpredictability can be debated: the logic of the electricity market implies that the price is very volatile, much more than what can be seen in other products, including oil. This price has already been negative many times in the past. Article 10 of the contract, as written, undoubtedly served as the basis for the court's decision given that it deviates from the traditional conception of *force majeure*.

It was the phrase "reasonable economic conditions" that led to debate in this dispute. The fundamental question is whether the clause of *force majeure* would apply automatically whenever "reasonable" economic conditions are no longer met. If this is the case, it would mean that we would no longer be using restrictive conception, but rather an extensive conception, of *force majeure*. This episode has only highlighted how urgently reform is needed for this ARENH mechanism, one established in a context where the market share of alternatives was modest.

The situation has changed significantly since, at the end of 2020, EDF lost more than a quarter of electricity consumption by the non-professional sector, and almost half of the electricity consumption by the professional sector. The historical operator still remains in a somewhat dominant position if we consider the French market alone, but that is no longer the case when we consider the European market (in particular the "Western Europe" copper plate formed by France, Germany, and the Benelux countries). The debate therefore focuses on the contours of what is called the *relevant market*, that is to say, the geographic area within which French consumers can find their suppliers.

2.7.2. *ARENH reform projects*

2.7.2.1. *The solutions proposed*

A controversy developed over whether or not it would be necessary to keep the ARENH market in place, or to modify the mechanism. Several solutions can be put forward:

– We could end this ARENH mechanism outright, considering that it has fulfilled its role and allowed many suppliers to enter the mass electricity market. The historic operator has lost a great deal of customers and it continues to lose more, even though its market share in the domestic market remains high. It is up to the parliament to decide, since this mechanism was created by a law. It all depends on the contours that are established for the *relevant market*. If we consider the French market exclusively, EDF is still in a dominant position (with more than 40% of the volume of electricity sold). If we consider the "Western Europe" plate (France, Germany and Benelux), EDF is no longer in a dominant position (with a market share close to 25%).

– The so-called "alternative" suppliers have called for an increase in the volume of the ARENH beyond 100 TWh per year (minimum 150 TWh), on the grounds that their customer numbers have increased and that this would allow them to mitigate the price increase for the end consumers by avoiding paying more for the "market supplement" (this of course assumes that the wholesale price is higher than the ARENH price). However, in a context where EDF is losing market share and where nuclear electricity production is trending downward and is set to decrease further given the objectives of the multiannual energy scheduling (with 50% nuclear by 2035), is it justified for the historic operator to continue to "subsidize" its competitors, especially when these competitors are currently oil companies and perhaps tomorrow digital companies such as GAFA whose revenue is much higher than that of EDF? This would be a kind of *death spiral*. The more they gain in market share, the more the "entrants" ask for an increase in the volumes of ARENH, which allows them to win new customers. Note that, given the surge in MWh prices on the wholesale electricity market at the end of 2021, EDF's competitors have obtained an additional 20 TWh of ARENH for 2022, but at an increased price of 46.2 euros/MWh.

– We can consider increasing the amount of the ARENH beyond 42 euros per MWh, in order to take into account additional costs related to "large-scale streamlining" and the increasingly tight safety requirements. The figure of 48 euros per MWh was mentioned by the CRE. This would counter an increase in the volume of ARENH.

– The ARENH could be designed to be a real option, and not a "free option" as is the case today: alternative electricity suppliers can currently benefit from the ARENH if they wish, but without supporting the long-term commitments related to the nuclear power plant network. For this, it would be necessary for a premium to be paid by the buyer of the option. With a premium, the ARENH would become an operation of common law.

– Finally, it may be considered that for the French, historical nuclear power has become an essential facility, the same consideration given to the electricity transmission and distribution networks. In other words, it is a kind of "common good", allowing to guarantee to all consumers a minimum base of electricity at a stable price that is not affected by the instability in the wholesale electricity market. In this case, at its limit, all nuclear power should be regulated in the same way as the other networks: the CRE could determine the level of the RAB (regulated asset base) corresponding to the value of the nuclear plant network and apply a normative rate of return to this value to determine the regulated price at which the nuclear MWh should be sold to all suppliers. These suppliers would include the "entrants", but also EDF.

2.7.2.2. *Nuclear power, an essential facility?*

Let us now look at the mechanisms and consequences of such a solution if in practice we opted for 100% ARENH, where all suppliers including EDF can subscribe to such a plan, but at a regulated price to be determined. The price could be fixed and based on the average cost of nuclear power. We would apply the same mechanism to nuclear power as that used to de-terminate the access charges to the networks, the RAB. The nuclear kWh could be sold on the market, but the operator would receive an additional income if the market price is lower than this target price, and it would pay any additional amounts if this market price is higher. This is the system used in the United Kingdom for the Hinkley Point nuclear power plant project C (Contracts for Differences [CfD]).

It is also possible to use a corridor mechanism, where the selling price on the market is limited by a ceiling price and a floor price, as shown in Figure 2.13. In the latter case, the sale price of the nuclear MWh would no longer be fixed, but it would vary within narrow limits (the Directorate General for Energy and the Climate [DGEC], at the Ministry of Ecological Transition, has called for a difference in the order of €6/MWh between the top and bottom values).

The corridor mechanism could also concern only a portion of the nuclear production, as envisaged by the CRE (for instance, 150 TWh out of a production in the order of 380–400 TWh).

This system recalls the European "monetary snake" which operated between 1972 and 1978, and it is very similar to the mechanism of "contracts for differences" established in the United Kingdom to finance the EPR at Hinkley Point, but with one difference: the final price is not fixed, but instead varies within narrow limits. With this system, the EDF-supplier becomes a common law buyer of French nuclear power when it sells electricity to its customers, and the company will have to respect a strict separation between its activities as a nuclear producer and those of an electricity supplier (this is the "blue-green" dichotomy of the "Hercules" project for the restructuring of the company currently under debate), particularly if the mechanism applies to the totality of nuclear production. The fact that all nuclear power will be sold on the market will increase its liquidity, whereas previously, the volume of ARENH was sold off-market at a regulated price.

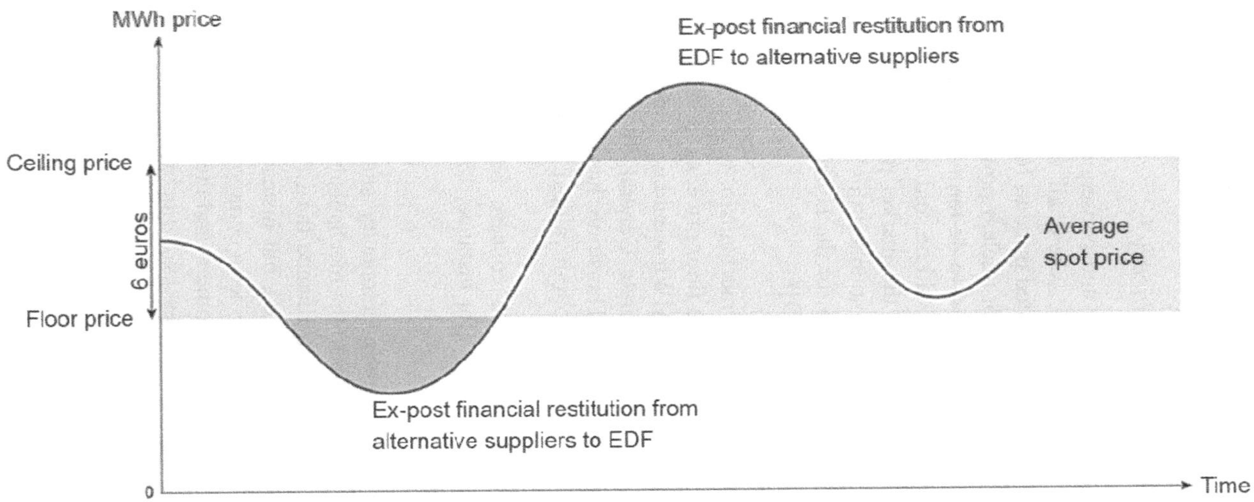

Figure 2.13. *The price corridor mechanism in the wholesale electricity market. For a color version of this figure, see www.iste.co.uk/percebois/nuclear2.zip*

The main advantage of the system is that this time the commitments between EDF and its competitors are symmetrical and no longer asymmetrical, as with the ARENH. This should avoid excess profits on both sides, both in the case that the spot price for EDF skyrockets, or conversely, when wholesale prices collapse and sometimes even become negative.

2.7.2.3. *Methodology for calculating the procurement cost of alternative suppliers*

We will present the total cost of supplying a given volume of electricity to alternative suppliers according to the ARENH system in force and the new corridor system as envisaged in the public consultation document of the DGEC. We will use C to represent the cost of supply for the alternatives (C_a with the current system, C_n with the new system) and P^* for the price of electricity on the wholesale market.

– *ARENH system in force*: within the framework of the current system, the ARENH ceiling Q' (currently 100 TWh) is acquired by the alternatives at a price P' fixed by the regulator (at 42€/MWh since 2012). If the demand for nuclear electricity from the alternatives Q_1 (133 TWh in 2019, 147 TWh in 2020 and 2021) exceeds the ceiling, they must acquire the additional $(Q_1 - Q')$ at the wholesale market price:

$$C_a = P'Q' + (Q_1 - Q')P^* \quad [2.3]$$

– *New corridor system*: the new system introduces a floor price P_{min} and a ceiling price P_{max} for the purchased nuclear TWh that is not for the TWh of nuclear energy purchased by alternative suppliers. Concretely, the alternatives buy their electricity at the market price and financial adjustments are made ex post if the average price on the futures market ($\bar{P}$) is established outside the ARENH price corridor.

The alternatives pay the market price for the acquired volumes and no financial compensation is made ex post if the average price that is determined is between the boundaries of the corridor:

$$C_n = P^*Q_1, \text{if } P_{min} \leq \bar{P} \leq P_{max} \quad [2.4]$$

The customers pay the market price and receive the difference at the end of the delivery period $(\bar{P} - P_{max})Q_1$ if the average price is higher than the ceiling price:

$$C_n = P^*Q_1 - (\bar{P} - P_{max})Q_1, \text{ if } \bar{P} > P_{max} \quad [2.5]$$

The alternatives pay the market price and pay the difference at the end of the delivery period $(P_{min} - \bar{P})Q_1$ if the average price is lower than the floor price:

$$C_n = P^*Q_1 + (P_{min} - \bar{P})Q_1, \text{ if } \bar{P} < P_{min} \quad [2.6]$$

2.7.2.3.1. Simulation

From the above methodology, we may simulate the potential effects of the ARENH reform. To do this, we use an ex post approach, observing what the procurement costs of alternative suppliers could have been under different reform options based on historical data.

Given the particular context of the year 2020 and the disruptions to the energy markets caused by the COVID-19 crisis, we have chosen to use the 2019 data for the calculations carried out in the reference scenario. In addition, we compare these results with those obtained from the prices of the 2018 and 2020 contracts for deliveries occurring in 2019 and 2021, respectively.

Within the framework of the system currently in force, the request for ARENH of the alter-natives in 2019 for the year 2020 was about 150 TWh, that is, 50 TWh higher than the ARENH ceiling of 100 TWh. The alternative suppliers therefore had to use the market for their complementary supply. If the corridor system had been in place, the alternatives would have bought all 150 TWh ordered on the wholesale market and would then have had to pay or receive additional remuneration depending on whether the annual average price on the futures market was lower than the floor price or higher than the ceiling price.

We take the futures prices of the contract for delivery in 2020 as a reference for the electricity market price and consider that the alternatives will smooth out their electricity supply over the year. This means we consider that they will make homogeneous purchases of the quantity Q_1 over all days of the year, and that the price P^* is equal to the annual average price $\bar{P}$ of the futures contract.

We also consider an alternative scenario ("ARENH revisited") in which the ARENH ceiling increases from 100 to 150 TWh in return for an increase in the regulated price for the sale of nuclear power beyond €42/MWh.

Based on various assumptions and from the previous ratings, the average cost of supplying the volume Q_1 under the different scenarios is given by Table 2.3.

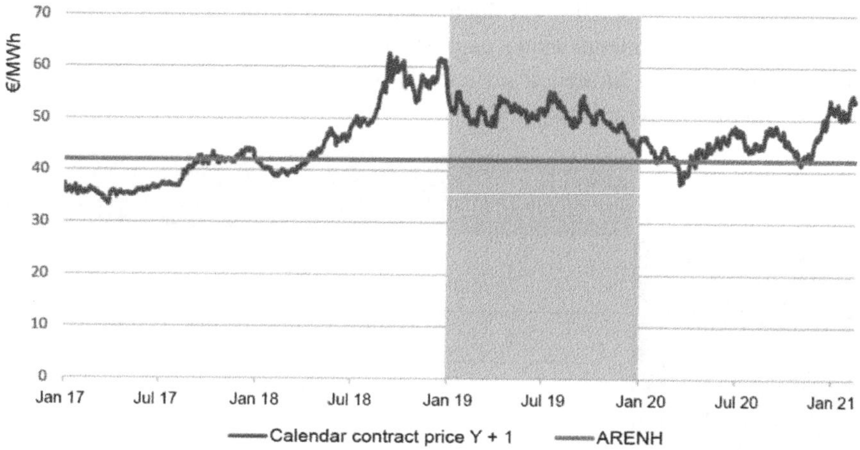

Figure 2.14. *Electricity prices in France for 1-year delivery (Y + 1). For a color version of this figure, see www.iste.co.uk/percebois/nuclear2.zip*

Scenarios	Equation for the average cost for a volume Q_1
Current (ARENH 42)	$P'(Q'/Q_1) + (Q_1 - Q'/Q_1)\bar{P}$
ARENH revisited	P'
Corridor	$\bar{P}$; P_{min} ; P_{max}
Market	$\bar{P}$

Table 2.3. *Equations of the average supply cost of the volume Q1*

In 2019, electricity to be delivered in 2020 traded on average at €51/MWh, a higher level than that of the ARENH, which explains why the volumes requested were much higher than the ceiling. The alternatives had to acquire the additional amounts on the market at an average price of €51/MWh, that is, a total supply cost of €45/MWh on average. The ARENH has therefore saved an average of €6/MWh on the 150 TWh acquired in 2020 (Figure 2.14).

The transition to a corridor system or a revision of the ARENH consisting of raising the transfer price and the ceiling should lead to a significant increase in the provision cost of about €45/MWh on a given day against more than € 50/MWh the following day, as can be seen in Table 2.4. The result will obviously depend on the choices that are made in terms of regulated prices or the price limits of the corridor. For the price levels considered below, we can see that the results are close, but are likely to vary depending on fluctuations in the future price. If it is desirable for the financial adjustments made ex-post between alternative suppliers and the historical producer to be limited, a sufficiently broad corridor should be provided, or the device should be revised frequently.

		2019 Prices (delivery in 2020)	2018 Prices (delivery in 2019)	2020 Prices (delivery in 2021)
Current system (ARENH 42 €/MWh)		45	44	43
Market		51	49	45
ARENH revisited		€48/MWh		
		€50/MWh		
		€52/MWh		
Corridor	44–50 €/MWh	50	49	45
	46–52 €/MWh	51	49	46
	48–54 €/MWh	51	49	48

Table 2.4. *Average cost of supply under different scenarios of reform and market prices*

The market only sets a small proportion of the price paid by the end consumer including VAT (in the case of a TRV household): based on the 2020 rates, taxes represented 33% of the price including VAT, tolls for access to transport and distribution networks approximately 32% and the supply balance (production and marketing of kWh) is 35%. This last part is where the competition and the wholesale market are likely to come into play. But part of the kWh that is commercialized is also sold at a regulated price, which is found in the so-called stacking mechanism. This is the case for a portion of the nuclear power, which corresponds to the ARENH (100 TWh out of 380 TWh of nuclear power produced, or about 18% of the total electricity production).

Note that if we add renewables, such as PV solar and wind, which benefit from guaranteed purchase prices (their share of production was about 8% or 43 TWh in 2019), there share ultimately becomes 25–26% of the total electricity production that can be considered to be regulated.

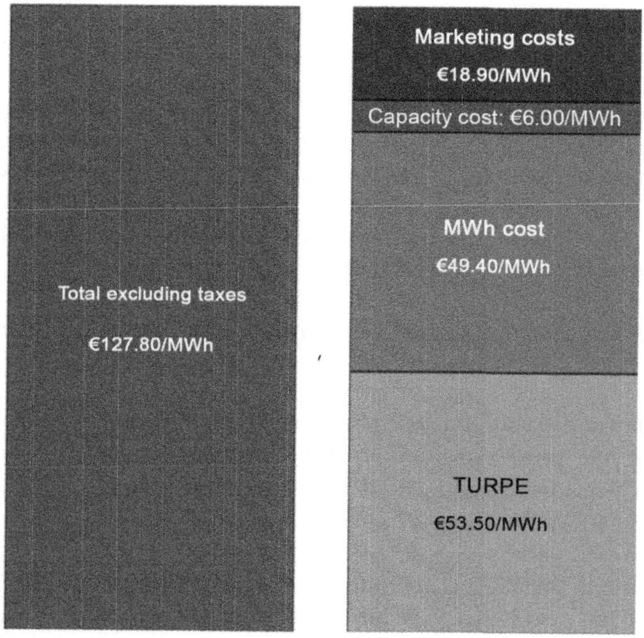

Figure 2.15. *Breakdown of the regulated sales tariffs excluding taxes for residential consumers as at February 1, 2021 (source: CRE data). For a color version of this figure, see www.iste.co.uk/percebois/nuclear2.zip*

The price per MWh, 193.5 euros, or 19.35 euro cents/kWh, is broken down as shown in Figure 2.15.

Excluding the effect of ARENH clipping (due to the achievement of the 100 TWh ceiling), ARENH volumes represent an average of 70% of the customers' consumption volume, with the remaining 30% (the supplement) being acquired by alternative suppliers on the market. Since 2019, the ARENH volumes requested by alternative suppliers are higher than the ARENH ceiling, which leads the CRE to proceed to cap these requests. For the year 2021, the allocation rate of the ARENH volumes requested by the alternatives was 68% (this corresponds to a capping rate of 32% for the alternatives, which requested 146.2 TWh of "ARENH" nuclear power for the year 2021, while the volume is capped at 100 TWh, or a rate equal to 100/146.2).

As a result, the quantity purchased on the market was more than 30% (which corresponds to the market complement of alternatives excluding the effect of the capping). The higher the capping rate, taking into account the volumes of ARENH ordered by the alternatives as well as the increased market complement of the alternatives, which requires increasing the weight of the market in the TRV in return to ensure *contestability*. In 2021, the market set only about 11% of the price including VAT paid by end consumers. The share is slightly higher for consumers who have left the TRV. On the wholesale market, the prices have been estimated on average at €46.6/MWh at the base rates and €60.6/MWh at the peak. The reference price for the additional capacity supply (excluding the ARENH clipping effect) is the average of the prices obtained during the six capacity auctions for 2021, that is, €31.241/MW. This increases the price per MWh by €6. The TURPE (*Tarif d'utilisation des réseaux publics d'électricité, transport et distribution*, "rate for the use of public electricity, transmission and distribution networks") is set by the CRE. The taxes are set by the Parliament.

Box 2.3. *Structure and change in the price of kWh paid by end consumers*

Evolution over the past 10 years: prices including VAT are increasing faster than tax-free costs: the price excluding VAT has increased by 40%, while the price including VAT has increased by 50% (consumer prices have only increased by a little more than 10%).

The increase in the TRV was limited to 4% in 2022 thanks to the "rate shield", when it should have exceeded 35% due to the rise in prices on the wholesale market.

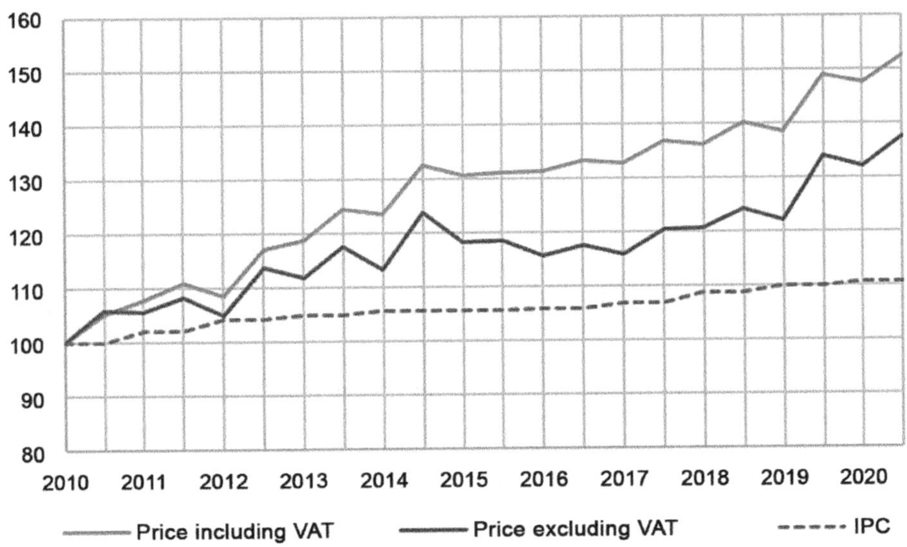

Figure 2.16. *Evolution of the price of electricity excluding VAT and including VAT for residential customers (with consumption between 2,500 and 5,000 kWh) and the Consumer Price Index (CPI, base 100 indices = 2010) (source: Eurostat and INSEE data). For a color version of this figure, see www.iste.co.uk/percebois/nuclear2.zip*

2.8. Conclusion

The competitiveness of historic nuclear power continues to be ensured, especially if we take into account the advantage provided by decarbonized energy and if we introduce the *system* costs related to the non-controllable nature of certain energies. This competitiveness is undoubtedly less secure if we look at the cost of new nuclear power, but again, this all depends on the price of carbon and the price of renewable energy storage.

Nuclear power is hardly compatible with the logic of an electricity market in which short-term prices fluctuate day by day. These are long-term investments that are poorly suited to a system that gives no guarantee as to the price at which the kWh of electricity will eventually be sold. At the prices observed in recent years on the wholesale electricity market, no investment in electricity production is profitable in Europe. The only investments made in the production of electricity are those that benefit from regulated prices, as is the case with solar and electricity.

The wholesale electricity markets are expected to be reformed as the share of renewables increases, since both renewables and nuclear power are similar in that the weight of their fixed costs in the cost price of kWh is much higher than that of their variable costs. Increasingly, the compensation for these investments will have to be based on power (kW) and not on energy (kWh).

The liberalization of the electricity sector hardly gives top priority to market mechanisms in setting the price of kWh, since the share of what is regulated (networks, taxes) is important, and the exceptions to the principle of competition (FIT, ARENH and TRV) are numerous. The market sets a small proportion of the price per kWh in France (including VAT) and the competition between suppliers actually focuses mainly on the marketing margins (which represent about 5–8% of the price including VAT).

The role that nuclear power will play in the future in France (and in Europe) will therefore largely depend on public choices as to the role given to regulation, and those given to competition. Maintaining or even renewing nuclear power capabilities will require more regulation and a long-term vision of national priorities. The renewal of the nuclear power network is based on a guaranteed price mechanism, similar to the one that renewables benefit from, and the same reasoning.

2.9. References

ACER (2020). Annual report on the results of monitoring the internal electricity and natural gas markets in 2019. European Union Agency for the Cooperation of Energy Regulators and the Council of European Energy Regulators.

Champsaur, P. (2009). Rapport de la commission sur l'organisation du marché de l'électricité. Ministère de l'Economie, Paris, France.

Cour des comptes (2012). Les coûts de la filière électronucléaire. Cour des comptes, Paris, France.

Cramton, P. and Stoft, S. (2006). The convergence of market designs for adequate generating capacity with special attention to the CAISO's resource adequacy problem. White Paper, Electricity Oversight Board, University of Maryland Libraries, MD, USA.

CRE (2020). Delibération N°2020-177 du 15 juillet relative à l'évaluation des charges de service public de l'énergie pour 2021 [Online]. Available at: https://www.cre.fr/Documents/Deliberations/Decision/evaluation-cspe-2021.

Cyclope (2021). *Les marchés mondiaux*. Economica, Paris, France.

Dambrine, F. (2019). Analyse microéconomique de l'intégration des EnR électriques intermittentes dans un système de production électrique. *Annales des Mines. Responsabilité et environnement*, 93, 7–14.

De Perthuis, C. and Solier, B. (2019). Le rôle du prix du carbone dans la transition électrique. *Annales des Mines. Responsabilité et environnement*, 93, 81–85.

European Commission (2020). Amended proposal for a regulation of the European Parliament and of the Council on establishing the framework for achieving climate neutrality and amending Regulation (EU) 2018/1999 (European Climate Law). European Commission, Brussels, Belgium.

Graham, P. (2018). Review of alternative methods for extending LCOE calculations to include balancing costs. CSIRO, Australia.

Hansen, J.-P. and Percebois, J. (2019). *Énergie : économie et politique*. De Boeck, Brussels.

IEA (2019). Electricity information. IEA, Paris, France.

IPCC (2006). IPCC guidelines for national greenhouse gas inventories. National Greenhouse Gas Inventories Programme.

Ministère de la Transition écologique et solidaire (2019). Stratégie française pour l'énergie et le climat. Programmation Pluriannuelle de l'énergie 2019-2023, 2024-2028. Ministère de la Transition écologique et solidaire, Paris, France.

Ministère de la Transition écologique et solidaire (2020). Nouvelle régulation économique du nucléaire existant. Consultation document, Ministère de la Transition écologique et solidaire, Paris, France.

Percebois, J. (2018). Promoting nuclear energy: Market pricing or regulated tariffs? *International Journal of Power and Energy Research*, 1(2), 81–92.

Percebois, J. and Pommeret, S. (2018a). Surcoût lié au stockage des EnR dans un mix électrique nucléaire à 50 %. *La revue de l'énergie*, July–August, 10–25.

Percebois, J. and Pommeret, S. (2018b). Cross-subsidies tied to the introduction of intermittent renewable electricity. An analysis based on a model of the French day-ahead market. *The Energy Journal*, 39(3), 245–268.

Percebois, J. and Pommeret, S. (2019). Storage cost induced by a large substitution of nuclear by intermittent renewable energies: The French case, *Energy Policy*, 135, 111067 [Online]. Available at: https://doi.org/10.1016/j.enpol.2019.111067.

Percebois, J. and Pommeret, S. (2021a). Les coûts cachés de l'électricité. Reflets de la Physique. SFP & CNRS, Paris.

Percebois, J. and Pommeret, S. (2021b). Efficiency and dependence in the European electricity transition, *Energy Policy*, 154, 112300 [Online]. Available at: https://doi.org/10.1016/j.enpol.2021.112300.

Percebois, J. and Solier, B. (2020a). Coronavirus : les conséquences sur le marché de l'électricité. *The Conversation*, March.

Percebois, J. and Solier, B. (2020b). La régulation du prix de l'énergie nucléaire en France : de l'ARENH au "corridor". Working document, hal-02942109.

Quemin, S. and Trotignon, R. (2021). Emissions trading with rolling horizons, *Journal of Economic Dynamics and Control*, 125, 104099 [Online]. Available at: https://doi.org/10.1016/j.jedc.2021.104099.

Roques, F., Verhaeghe, C., Le Thieis, Y., Li, Y. (2016). Électricité : tous les impacts du mécanisme de capacité français. *Revue de l'énergie*, 634, 41–51.

RTE (2021). Bilan électrique 2020. RTE, Paris, France.

SDES (2020). Chiffres clés de l'énergie. Ministère de la Transition écologique, Paris, France.

Stoft, S. (2002). *Power System Economics: Designing Markets for Electricity*. Wiley-IEEE Press, Piscataway, NJ, USA.

Timilsina, G. (2020). Demystifying the costs of electricity generation technologies. Policy Research Working Paper, World Bank, Washington, D.C., USA.

3

The Industrial Challenges of Nuclear Power

Xavier URSAT
Groupement des industriels français de l'énergie nucléaire, Paris, France

The French nuclear industry has reached a decisive stage in its history. It is faced with the challenge of carrying out the nuclear recovery program initiated by the President of the Republic in February 2022, and more generally, taking part in the global dynamics in favor of nuclear energy at a time when it is becoming crucial to contribute to the decarbonization of the economy, and asserting itself as an effective vector for industrial recovery.

The energy transition is well underway worldwide, while the population of the planet will continue to grow and its standards of living will improve, leading to an increase in global electricity consumption in the order of 50%. However, given the uncertainties as to the extension of reactors in some OECD countries, 25% of the world's nuclear capacities could be discontinued by 2025, with this figure even reaching two-thirds by 2040. Without any new construction, and given the technologies currently available, the world economy could then generate an additional 4 billion tons of CO_2, mainly by an increased use of natural gas made necessary to fill the gaps in energy provision that are inherent to renewable energies.

The markets for new nuclear power are driven by this increased awareness about climate issues. Nearly 110 countries are now seeking to achieve carbon neutrality by 2050, with strategies that include adding new nuclear production capacities into their mix of electric power sources. This choice to use nuclear energy is recommended in all IPCC and IEA scenarios; some of their projections even call for the current installed nuclear capacity to be increased six times over.

France is one step ahead, and many countries are now looking to benefit from four key assets found in the country's mix of electricity sources: decarbonized production[1], energy independence, the local industrial sector and competitive electricity production. To preserve these assets and to fulfil its commitment to achieving carbon neutrality, several scenarios have been projected, but the most realistic rely on nuclear power to form a stable base alongside renewable energies, which includes building new reactors within France.

The French nuclear industry, the well-established cornerstone of the implementation of this "electric future", is ready to accept the challenge and has given itself the means to do so by committing to a vast, in-depth plan to transform its methods, tools and its culture. Now, the confidence and support the government has placed in this plan have allowed the achievements arising from this new momentum to take root.

3.1. Nuclear power, part of the energy transition

In nearly 50 years, France has developed an industrial sector for carrying out all operations related to the nuclear fuel cycle: the extraction, conversion and enrichment of uranium, the manufacturing of fuel, the treatment of spent fuel and the reuse of materials resulting from this treatment, as well as the packaging and storage of waste downstream of the cycle. In this way, French industry has mastered the know-how involved in these processes.

3.1.1. *Nuclear power, the third largest industrial sector in France*

After the automotive and aerospace sectors, nuclear power presents an opportunity for the French economy, since promoting the decarbonization of industrial activity; it carries with it part of the dynamism of the country.

Nuclear power is a sector with stable and long-term activity. Its power plants, equipment manufacturing plants, and research centers are rooted in France's territories in particular[2] and allow for the development of infrastructures and, more

1 The average total emissions are less than 50 g/kWh, compared to 400 g/kWh in Germany and 260 g/kWh in Italy.

2 Source: mapping of the nuclear sector (CSFN).

broadly, of an entire industrial ecosystem that benefits other economic sectors. The sector represents almost 7% of industrial employment, totaling more than 220,000 non-relocatable jobs[3]. Almost half of the sector's workforce is concentrated in Île-de-France, the Auvergne-Rhône-Alpes region and Normandy and are also distributed throughout the rest of the country. The qualification levels are somewhat higher, with 35% of managers and 43% of technical and specialized employees. Figure 3.1 summarizes these data.

The segments are as follows:

– sanitation-deconstruction;

– waste management;

– boiler construction, foundries and forging;

– air conditioning and ventilation;

– control-command;

– electricity, digital technology and instrumentation;

– tests and controls;

– energy output;

– generators, turbo-alternator groups;

– civil engineering, first-level equipment support;

– engineering, intellectual services and technical assistance,

– logistics, transport, lifting and handling;

– rotating machines;

– nuclear process;

– radiation protection;

– site protection;

– valve installations;

– water treatment;

– welding of pipes.

Box 3.1. *The 19 segments of the activities of the nuclear industrial sector*

3 Source: mapping of the nuclear sector (CSFN).

There is significant potential for development. The nuclear sector recruited nearly 30,000 employees over the period of 2015–2018 – almost double the projected number announced by the companies in the sector in 2014 – and expects more than 22,500 hirings over the period of 2020–2023, driven by major nuclear projects committed to be carried out on our territory or being developed internationally, of which the most important are the EPR.

This is also an activity that produces exports. French industrialists in the sector work on nearly 250 reactors around the world and are regularly called on by electricity providers wishing to develop a nuclear program. In fact, the French industry – led by EDF, whose experience as the top global nuclear operator is a proven competitive advantage – is capable of designing a power plant in its entirety, from the "nuclear island" to the conventional section, but also offering a range of services throughout the entire value chain, from finalizing the design of the power plant to its dismantling, including its construction and operation/maintenance.

* Scope of study: all stages contributing to the nuclear cycle process, from R&D and studies in advance to waste treatment (excluding mining extraction), centered on the production of civil nuclear energy.

Figure 3.1. *Summary of data for companies in the nuclear sector (source: CSFN 2019 survey – Strategic Committee on the nuclear sector). For a color version of this figure, see www.iste.co.uk/percebois/nuclear2.zip*

3.1.2. *The EPR, a nuclear technology of the future*

The EPR is the flagship product offered to countries with significant electricity production needs and with power grids scaled for high power levels (SFEN 2021). Authorized by four safety authorities, it adheres to the highest safety standards and offers optimized technical and environmental performance (less fuel used with constant electricity, reduction of the duration of shutdowns for recharging, reduction of radioactive waste and effluents, etc.).

Because of the successful entry into service of two reactors in Taishan (China) and the Olkiluoto 3 reactor (Finland) and to the experience gained from Flamanville 3 and the two reactors at Hinkley Point C (United Kingdom), the construction capability and competitiveness of the EPR are continuously improving.

And the EPR network may grow even larger, as two EPR projects are already under development:

– The Sizewell C project, which provides for the construction of two EPRs (3.2 GW) on the east coast of England, is currently being studied by the British government, which at the end of 2020 announced that it had entered into discussions with EDF Energy to reach a financing agreement.

– In India, in April 2021, the EDF group submitted a binding final offer to NPCIL, the Indian public energy company with which the company has been in exclusive negotiations since the end of 2018, for the sale of six EPR reactors (~10 GW in total) to be installed at the Jaitapur site.

In addition, particular attention has been paid to developments in the European nuclear market, in particular in Poland, the Czech Republic, Slovenia and the Netherlands, where the prospects for the development of nuclear energy, buoyed by the ambitions of the European Green Deal, are promising.

All these projects will make it possible to develop synergies with the first EPR put into service, while contributing to the maintenance of the skills of the French and European nuclear industry. Together with the Finnish company TVO, EDF and its French-Chinese subsidiary TNPJVC are also the founding members of the group of EPR reactor operators, the EPR Owners-Operators Group (EPROOG).

3.1.3. *NUWARD, the French project for SMRs*

Small modular reactors (SMRs) are low-power reactors, of less than 300 MWe per module, designed to be manufactured in factories using modular and standardized processes, with shortened construction times.

With their experience, the engineering and the industrial fabric that make them up, EDF, the CEA, Technicatome and Naval Group have now undertaken the development of a model for an SMR:NUWARD, on the principle of a pair of 170 MWe units. This new reactor will make it possible to support countries wishing to accelerate the decarbonization of their energy grids, namely, by replacing their oldest coal-fired or oil-fired power plants with new low-carbon production capacities.

The development roadmap of the NUWARD project involves a summary preliminary project phase over the period of 2019–2022.

When this PSA phase is completed, the safety options file (DOS) of the project will be filed with the Nuclear Safety Authority (ASN). From 2023 to 2026, the basic design phase of the project (including the complete design of the power plant, equipment specifications, confirmation of architectural choices, real tests of certain systems, development of calculation tools and safety demonstrations) will be carried out.

Beginning in 2026, the phase of detailed design will start if the decision to build a reference power plant is officially taken and enacted through regulation, and the investment decision is formally acted upon.

EDF and its partners aim to build a reference power plant in France by 2030, which will confirm the constructability and performance of the product and establish the technology vis-à-vis the control authorities.

3.2. A sector committed to a dynamic of deep transformation

The nuclear sector is now a major sector within the French economy. But to stay ahead, it is essential to strengthen its industrial performance and competitiveness. The nuclear sector is equipping itself with the means to do this by restructuring itself, transforming its working methods and tools, and opening up to its industrial ecosystem. And next on the horizon is the extension of the existing plant network and the construction of the main one in France and also internationally.

3.2.1. *The restructuring of the nuclear sector since 2015*

Between 2015 and 2018, the French nuclear sector was restructured to ensure the continuity of France's ability to pursue its nuclear policy. Its objectives is to clarify the respective roles of the various actors, simplify processes and optimize systems.

In 2016, the EDF Group became the majority shareholder of Framatome, from the Areva branch responsible for activities relating to nuclear reactors (75.5% of the capital). The two companies have joined their engineering teams in the design and construction of the nuclear island in a joint subsidiary: Edvance. The Areva branch in charge of nuclear fuel cycle activities became Orano, and Areva TA, in charge of the engineering and construction of research reactors, became Technicatome.

The French nuclear sector now relies on actors with clearly defined responsibilities:

– *EDF*: is an owner and operator of the French nuclear power plant network, a leader of engineering over the entire lifecycle of its nuclear reactors (design, construction, maintenance, and operational improvements, deconstruction), and a leader in the development of international partnerships for the realization of nuclear reactor construction programs.

– *Framatoma*: is a subsidiary of the EDF Group, overseeing the design and manufacturing of the nuclear boiler equipment and manufacturing the fuel assemblies. This industrial organization of the "reactors" section is completed to cover the entire cycle.

– *Orano*: is an owner and operator of fuel cycle facilities:

- upstream activities: uranium exploration and mining; strategic stocks in uranium, conversion and enrichment of uranium;

- downstream activities: uranium recycling, manufacturing of MOX fuel and vitrification of high-level wastes from fuel.

– *ANDRA* (*Agence nationale des déchets radioactifs*, the French National Agency for Radioactive Waste): includes management of the different categories of radioactive waste from electricity production, as well as from all other uses of nuclear energy in France (medical, etc.).

– *CEA* (*Commissariat à l'énergie atomique et aux énergies alternatives*, the French Commissariat for Atomic Energy and Alternative Energies): undertakes research and development of nuclear power, owners and operates several research reactors, including the Jules Horowitz reactor currently under construction.

In addition, the industrialists of the nuclear sector have established a professional union in 2018 through creating the French Nuclear Energy Industrialists Group (*Groupement des industriels français de l'énergie nucléaire* [GIFEN]). This group brings together more than 200 member companies, as well as similar entities such as the France-China Electricity Partnership association (*Partenariat France-Chine électricité* [PFCE]), which supports French small and mid-size businesses in their exporting activities in China; the International Institute of Nuclear Energy (*Institut international de l'énergie nucléaire* [I2EN]), which provides services offered from French providers in terms of training for personnel in the nuclear sector; the French

Association for the Regulations of Design, Construction and Operational Monitoring of Nuclear Boiler Equipment (*Association française pour les règles de conception, de construction et de surveillance en exploitation des matériels des chaudières électronucléaires* [AFCEN]); the Bureau for Standardization of Nuclear Equipment (*Bureau de normalisation d'équipements nucléaires* [BNEN]); the French Society of Nuclear Energy (*Société française d'énergie nucléaire* [SFEN]); and "Nuclear Valley", designated as a competitiveness cluster for the nuclear sector.

Within the GIFEN, large companies, mid-sized companies and small companies conduct joint projects within the framework of eight commissions dedicated, respectively, to prospects in France and the offers of the sector, nuclear safety and quality, digital, skills and training, European affairs, international, innovation, and R&D and communication.

It is also notable that the French nuclear sector is one of the sectors considered to be strategic by the French public authorities. The Strategic Nuclear Sector Committee (*Comité stratégique de filière nucléaire* [CSFN]), established in January 2019 within the framework of the National Industry Council (*Conseil national de l'industrie* [CNI]) chaired by the Prime Minister, brings together the main actors and stakeholders of the sector: major contractors and major industrialists, representatives of small/midsize businesses making up the fabric of the nuclear industry, trade union organizations representative of the sector, professional associations (FIM and SERCE) as well as the federal government, through the two relevant ministries: the Ministry of Economy and Finance, via the General Directorate of Enterprises (*Direction générale des entreprises* [DGE]), and the Ministry of Ecological and solidarity transition, via the General Directorate of Energy and Climate (*Direction générale de l'énergie et du climat* [DGEC]).

The organization established with the CSFN and the GIFEN has allowed the sector to benefit from a simplified organization involving all the actors: the State, the major contractors and the entire industrial fabric.

This reorganization was intended to contribute to revitalizing the performance of the sector by restoring clarity to the driving forces of nuclear power in France, facing the industrial difficulties of the first EPR sites, which they still have to address, in particular at the Flamanville construction site. A thorough diagnosis was then carried out to build a robust and complete transformation plan.

3.2.2. *The unfavorable industrial context during the first days of the EPR*

An analysis of the reasons for the difficulties encountered was made in 2019 by the industrialist Jean-Martin Folz (2019) in his report on Flamanville 3, and by the GIFEN on all EPR projects in the world (GIFEN 2019a). EDF and the rest of the nuclear industry have learned lessons from these analyses (GIFEN 2019b, 2020) and initiated performance plans to continue strengthening their ability to perfect the construction of nuclear reactors at the highest level of performance.

The initial objective assigned to the first EPR in France was to develop the new reactor model, its technical and regulatory framework, and to retain the skill sets needed to carry it out. The project was a transfer of industrial resources in anticipation of a future works as part of a likely renewal of the French nuclear plant network. Furthermore, it was under these terms that the decision was taken “to keep open the option of nuclear power”. Nevertheless, several contextual elements led to the launch of the project, with the targets for cost and construction duration underestimating the needs of the mission.

In particular, the Flamanville 3 project had to overcome the deterioration of available skills among all the players directly or indirectly involved, after a long period of inactivity in nuclear construction in France. In fact, more than 20 years had passed since the lead unit of the N4 series (Chooz B1) and the EPR at Flamanville. The generational change during this period, and the decline in levels of practical experience that resulted from it, disrupted the perception of this loss of skills, the appreciation of the sources of risks and the control of the performance factors that had made the historic success of the French nuclear program.

In addition, the context in the beginning of the 2000s was marked by profound transformations: the deindustrialization of the French economy, increased commercial competition on the international nuclear market, the recomposition of the nuclear sector at the end of the national equipment program and the contractualization of societal relations. All of these elements weighed on the launch and subsequent execution of the Flamanville 3 project.

This led to, for instance, the accumulation of production errors in civil engineering, the welding of pipes and the manufacturing of equipment, which led to a high rate of activities that had to be repeated and for controls to be multiplied, at the cost of an increase in the workload and greater complexity in management decisions.

The production of the Olkiluoto EPR in Finland, under the leadership of a consortium between Areva and Siemens for the operator TVO, has similarities with the Flamanville experiment and confirms the lessons that were learned. The two Chinese EPRs, commissioned in 2018 and 2019 in Taishan, also offer valuable experiences. By starting ahead of the AP1000 competitor, after 9 years of construction and at a cost generally consistent with the initial estimate by EDF, despite the transition to a lead unit situation during the project, Taishan is a success that reinforces the analyses and lessons learned from the execution of the first EPR.

3.2.3. *The Excell plan: a return to excellence*

Based on these learnings, and in connection with the GIFEN, the Excell plan (Figure 3.2) announced by EDF in December 2019 seeks to enable the sector to achieve the highest level of rigor, quality and excellence to be the priority in major nuclear projects:

> With the Excell plan, we want to achieve rapid results, in all companies and factories in the nuclear sector. Our objective is to prioritize our major projects, current and future, in France, the United Kingdom and elsewhere in the world, allowing nuclear energy to contribute to the fight against climate change to the fullest. (Jean-Bernard Lévy)[4]

In April 2020, the management of this plan was entrusted to a general delegate for industrial quality and nuclear skills, affiliated with the CEO of EDF.

The plan made 25 commitments, divided into five priority areas of work:

– State-of-the-art project management, including the implementation of oversight for major "new nuclear" projects to guarantee each benchmark is met.

– A development of the skill sets in the French nuclear sector, and in particular of 21,000 professionals who will join it between 2019 and 2022, with the support of the French government's national investment program "France relaunches".

– The manufacturing and construction are guaranteed to be compliant on the first production, with the objective of quickly detecting and addressing compliance issues by deploying "zero defect" action plans in each factory in the sector.

4 This is an EDF press release from Thursday, October 15, 2020. EDF and the nuclear sector present the first stage of the Excell plan: https://www.edf.fr/groupe-edf/espaces-dedies/journalistes/all-press-releases/edf-presents-a-first-point-of-detail-of-the-plan-excell-the-plan-dexcellence-de-la-filiere-nucléaire.

– A supplier relationship based on simpler contracts and focused on results. The actions are committed under the "France relaunches" plan to consolidate the sector.

– The strengthening of quality and safety through standardization and replication, so as to ensure cost estimates and deadlines are met. Standardization is a major driver, both for large and small production series. All companies entering within the framework of the Excell plan (Alstom Transport, Airbus satellite, Naval Group, etc.), including competitors in the French sector (Rosatom, CGN, etc.), confirm this and work toward gains in the order of 30%. For industrialists, standardization is also the starting point for the renewal of designs and the repetition of production, guarantees of quality and safety.

In terms of performance, the Excell plan aims to:

– reduce the recoveries of studies and manufacturing for the first pair of EPR2 reactors to one-tenth the current level or less (compared to the EPR from Flamanville);

– secure the expected time and cost gains for Sizewell C and the first pair of EPR2 reactors (compared to the EPR at Hinkley Point C);

– permanently regain optimum skill levels for welding operations on nuclear construction sites and on operating reactors (welding plan).

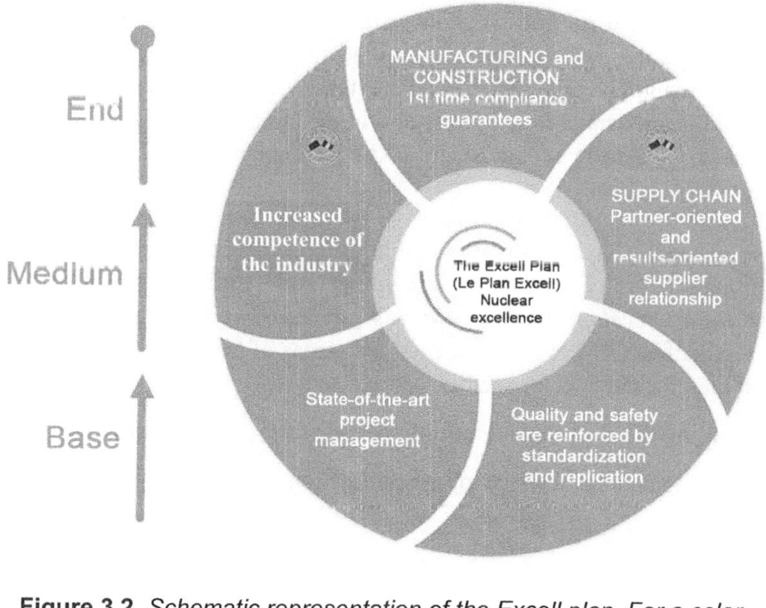

Figure 3.2. *Schematic representation of the Excell plan. For a color version of this figure, see www.iste.co.uk/percebois/nuclear2.zip*

At the end of 2021, 23 of the 25 commitments of the Excell plan had reached or even exceeded its target levels. The execution involves all the axes of the plan. Forty-four of EDF's suppliers have also launched their own plan, *Excell in quality*.

The third phase of the Excell plan is now underway, focused on an objective: consolidating the results achieved and continuing the actions undertaken to reach the best industrial standards.

3.2.4. *An effective dynamic*

The implementation of the Excell plan is already producing tangible results on current projects. Together with the substantive actions that affect the organization and the cross-cutting processes to make them more efficient, the commitments are intended in particular to strengthen the control of the production of tomorrow's reactors. This will allow for the detailed design phase with suppliers to be undertaken earlier and on a stronger basis than has been done previously. To demonstrate this, a few pertinent examples may be cited here.

3.2.4.1. *The maintenance and strengthening of local skill sets*

The lessons learned from the transition between the N4 housing and the EPR make it clear that part of the essential abilities needed for the launch of a new project quickly decline when no construction programs are underway. In addition, other elements such as the evolution of age groups, and the ability of players to acquire, develop and maintain new skills must be taken into account to ensure that the necessary skills are available in quality and in the appropriate amount when they are needed. Since 2021, the sector has used two complementary tools to meet these challenges.

These elements have made it possible to provide an orientation for studying skills through a project by the Employment and Ability Development Agency (*engagement de développement de l'emploi et des compétences* [EDEC]) from the nuclear sector. The GIFEN has identified five areas of action, which are included in the action plan for the nuclear sector:

- identifying the skill set and training needs of the nuclear sector;
- developing courses for providing certifications;
- developing learning;
- developing the attractiveness of the sector and related communications actions;

– supporting companies in the sector in taking into account the impact of digital technology on professions.

Alongside the EDEC, the creation of a University of Nuclear professions in April 2021 has made it possible to bring together the players within the nuclear sector and territorial actors in order to revitalize training mechanisms and to provide clarity and coherence to a complex ecosystem. It works as an "outside the walls" university, inspired by the feedback from other industries, in particular the naval industry, which created the Naval Industries Campus (Campus des industries navales [CINAV]) in 2019.

3.2.4.2. *ISO 19443, complementing ISO 9001 for the nuclear industrial chain*

The standard ISO 19443 defines "the specific requirements for the application of ISO 9001:2015 by organizations in the supply chain of the nuclear energy sector that provide products or services that are important for nuclear safety". It therefore applies a generic reference framework to a particular activity, as is already the case for other high-tech sectors (aerospace, rail, automotive, outer space, etc.). It has been chosen as a reference standard for all manufacturers in the sector, which allows for the standards to be harmonized, and therefore provides greater ease for manufacturers in the procedures related to it. As an international standard, it will also provide French manufacturers with better recognition in the countries where French industry expects to develop.

3.2.4.3. *A welding plan for strategic abilities throughout the entire industrial sector*

Taking into account the specific challenges in terms of skills and quality, the welding plan has been integrated into the Excell plan. In particular, this plan makes it possible to support the training and skill development of welders who work on sites for use by the nuclear sector.

The goal of the welding plan is to "weld it right the first time" by working both on skill sets, but also on the entire process of welding and on its environment. This particularly focuses on:

– The creation with the industry of standards for the mastery welding during the different stages of the process: a system of authorization for nuclear welders defined with the profession, the preparation and identification of risks in realization, the supervision of construction sites and post-execution controls.

– The creation of the Cotentin Superior Welding School (*Haute école de formation soudage du Cotentin*) (Hefaïs), in partnership with Orano, Naval Group and the CMN, which has been operational since late 2022.

3.2.4.4. *Digital transformation*

Digitalization is an essential element for the success of projects and the performance of the sector over time. In particular, the organization of information exchanges between all the actors involved in the same project is an important tool for shortening the time needed to design, manufacture and control components, increasing the availability of the plants, lowering operating costs and improving security.

Spurred by the major contractors – ANDRA, CEA, EDF, Framatome and Orano – the Genesis project thus aims to accelerate the digital transition of the nuclear sector. This requires in particular the provision of digital services within collaborative platforms intended for all players in the sector of any size (mid-sized companies, small and very small companies). These services include:

– the digitalization of the manufacturing process of equipment subject to ESPN (*équipement sous pression nucléaire*, nuclear equipment subject to pressure) regulations, from design to factory inspection, with key estimated gains of 20% on conformity certification costs, greater control of schedules, easier access to ESPN markets;

– the simplification and securing of data exchanges between the operator and the engineering companies involved in the modifications of the nuclear fleet, so that the entire supply chain increases in productivity.

3.3. Giving visibility to the sector through a long-term energy strategy

> What we have to build today [...] is the rebirth of the French nuclear sector. (Emmanuel Macron, President of France, Belfort, February 10, 2022)

The industry is now ready to tackle the challenge of created renovated and renewed network of nuclear power plants.

The mobilization of this industry, which has allowed it to accelerate its recovery in recent years, has now been strengthened further thanks to the visibility it has received from the French government. The government's decisions, which have been essential for building the future of the French and European electricity system, will allow EDF and the sector to invest in industrial equipment, instill skill sets and know-how, and thus maintain the ability to execute a large-scale program.

3.3.1. *In France: ensuring supply capacity and guaranteeing carbon neutrality*

On February 10, 2022, as part of a speech on the "Energies" component of France 2030, the President of the Republic unveiled three projects to be implemented in the country by 2050 to achieve carbon neutrality, reduce the country's energy dependence, and ensure industrial development:

– reducing France's energy consumption by 40%;

– accelerating the development of renewable energies;

– consolidating the role of nuclear power in France's energy mix.

Indeed, no one mode of production alone will be able to satisfy the electrical needs of tomorrow. Studies expect the share of electricity in end energy consumption to double: compared to 473 terawatt hours (TWh) of electricity in 2019, by 2050 France is expected to need 630 TWh, or possibly more. Nuclear power, which currently allows France to generate electricity without emitting CO_2, is competitive and available at all times; therefore, it will continue to remain essential.

3.3.1.1. *The extension of the operating period of the network: a path forward in the energy transition*

The French nuclear plant network now accounts for 67% of electricity production, with 56 nuclear reactors in operation. The average plant age depends on the technological level, from 19 years (for the so-called "N4" level) to 38 years (for the so-called "900 MW" level). Three quarters of the 60 GW of nuclear capacity currently in operation have reached 50 years by 2040.

That is why EDF launched the "Great Streamlining" program in 2014, an ambitious program that brings together 19 projects with the ability to respond to the extension of the exploitation of French nuclear power plants for a financial envelope of approximately €45 billion$_{2013}$ (i.e. 48 billion euros at the present) over the period 2014–2025. This ambitious program aims to:

– allow the operation of the reactors after 40 years;

– allow the network to achieve its production objectives in complete safety by controlling the duration of offline maintenance periods for only the time that is needed;

– secure and optimize the financial trajectory of investments.

The continuation of the operational lifespan of all the nuclear reactors "which can be operated" in the operating fleet, extending it beyond 50 years[5], will make it possible to benefit for at least another 10 or 20 years from the production of a network built between the years of 1970 and 2000 and which represents the most competitive base for carbon neutrality.

3.3.1.2. *The construction of new reactors: a necessity*

In addition to the extension of the operating period of the nuclear plants, the President announced the launch of a historic program for the construction of new nuclear reactors in phases.

This program is based on the EPR2, an optimized version of the EPR, forming part of the industrial continuity of the EPR. It maintains its assets while integrating the insights obtained from the projects carried out in France and around the world (China, Great Britain, Finland) and the operation of the French nuclear network. Priority is given to improving the ability to construct the reactor and its industrialization. The President thus expressed a desire to see the construction of six EPR2s initiated, and for studies on the construction of eight additional EPR2s to be launched.

EDF will build and operate the new EPR2. The construction site of the first EPR2 pair is set to begin in 2024. The civil engineering works for the reactor building will begin in 2028 (for the first reactor), with the two reactors being commissioned in 2035/2037.

The head of state also announced a call for sustained projects valued at 1 billion euros by France 2030 to support the development of the SMR project NUWARD developed by EDF and its partners to help bring to light innovative reactors that will make it possible to close the fuel cycle and produce less waste.

This major, unprecedented program could thus lead to the commissioning of 25 GW of new nuclear capacities by 2050.

It offers the assurance of achieving our strategic objectives with greater certainty and at a lower cost. This is what the RTE studies show: these scenarios that include new nuclear facilities are at least 15% less expensive than the scenarios without nuclear power (i.e. 10 billion euros in annual savings in 2060)[6].

---

5 This extension will be subject to a study by EDF of the conditions of this extension in conjunction with the French Nuclear Safety Authority (*Autorité de sûreté nucléaire* [ASN]). It will be carried out as part of a fifth periodic review, which will be instructed by the ASN.

6 Futurs énergétiques 2050, October 2021.

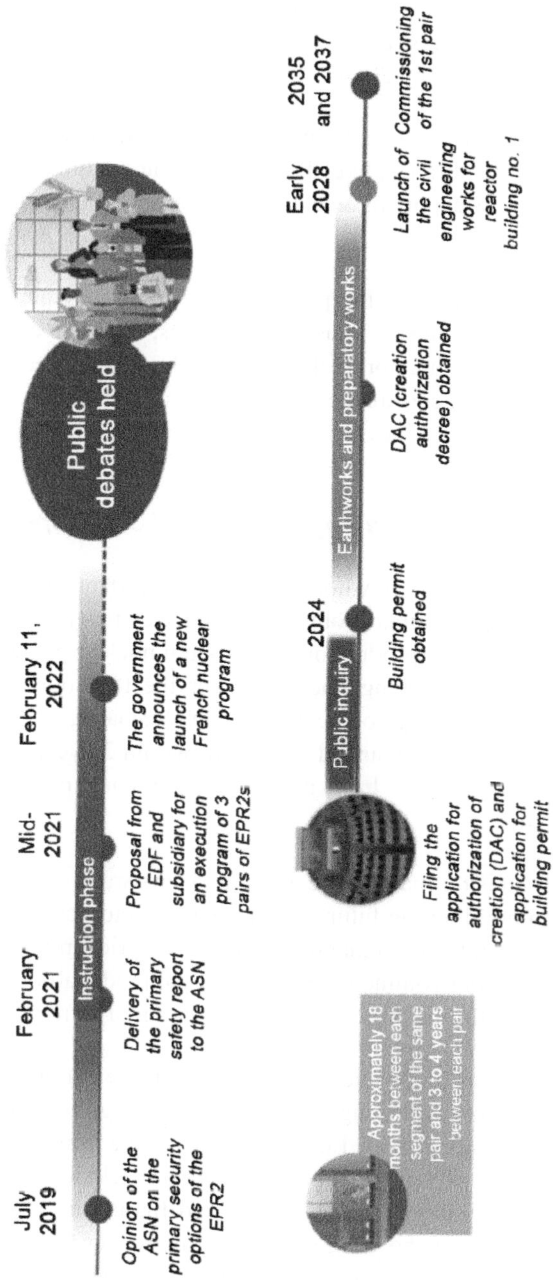

Figure 3.3. *EDF's provisional timetable for the construction of a first pair of EPR*

This dynamic will further strengthen French industrial capacities so as to obtain the necessary levers to build nuclear reactors in complete autonomy and at the best level of performance when the strategy is finalized in 10–15 years that will ensure the electricity supply of France by 2050.

It is ultimately a strong argument for establishing the legitimacy of France's industry outside its borders. France, like all other "nuclear manufacturing" countries, only has the legitimacy to export its nuclear know-how if it has the support of its government.

To implement the decisions of the president, the regulatory and organizational conditions of the sector and the state must be met. Thus, it will be essential to create an inter-ministerial program directorate dedicated to the new nuclear power plant to ensure the management, coordinate administrative procedures and ensure compliance with costs and deadlines for construction sites.

3.3.2. *Designing an energy strategy covering all of Europe*

Just like France, Europe as a whole is facing important challenges, such as securing the energy supply, climate objectives and electricity costs. Currently, a quarter of European electricity production comes from nuclear power and 13 out of 27 countries have reactors, making nuclear power the top source of low-carbon electricity in Europe. According to the IEA's SDS scenario, a volume of about 50 GW of new nuclear power is planned to be built within 20 years in Europe. Nine countries are already building or have initiated construction projects, including a "new entrant" (Poland).

The continent's decision-makers are now at a crossroads and must take the decisions essential to building the future of the energy sector, and more specifically the European electric sector. The signal quality of a CO_2 price in Europe, which is to say, its consistency with the volume objectives and the predictability of the changes it undergoes, is essential for the effective decarbonization of European production.

In this sense, the inclusion of nuclear power within the European taxonomy on February 2, 2022 is a political turning point. This is a recognition, albeit one that is late and imperfect, explained by a European text on the contribution of nuclear power to the fight against climate change. If it were to exclude nuclear power from the energy mix, the Commission would have risked finding itself at odds with the decarbonization ambitions of the Green Deal, and moreover, favoring investments from outside Europe to the detriment of European private and public financing, seriously weakening European industry and its historical know-how.

3.4. Conclusion

By clearly positioning nuclear power among the technologies of the future, France can envision an industrial and economic revival that is consistent with its climate objectives.

This choice that has been made by the President is historic and gives the nuclear sector a great responsibility. To respond to this, the sector now offers a technically perfected and efficient solution and is ready to launch the construction of several pairs of EPR2s.

This is our commitment. The stakes are high, since without a new program for the French nuclear industry, France would lose an important comparative advantage, and European states wishing to develop nuclear power would have no choice but to work with the American, Chinese or Russian sectors.

But securing the asset of nuclear energy over the long-term requires remunerating the capital invested in the tools for its production to a degree that is fair. It is urgent to implement true economic regulation of nuclear power and to integrate industrial issues into energy strategies in the long-term.

3.5. References

Folz, J.M. (2019). Rapport sur la construction de l'EPR de Flamanville. Ministère de l'Économie et des Finances, Paris, France.

GIFEN (2019a). État des lieux de la filière nucléaire française et retour d'expérience des projets EPR. GIFEN, Paris, France.

GIFEN (2019b). Conditions à remplir pour mobiliser les capacités industrielles nécessaires au renouvellement du parc électronucléaire. GIFEN, Paris, France.

GIFEN (2020). Plan d'action détaillé de la mobilisation des compétences et capacités industrielles pour le Nouveau Nucléaire en France. GIFEN, Paris, France.

SFEN (2021). Qu'est-ce qu'un réacteur EPR ? SFEN, France.

4

The Economic Costs of a Nuclear Accident

Ludivine PASCUCCI-CAHEN
Laboratoire d'analyse économique des risques nucléaires, IRSN, Fontenay-aux-Roses, France

The study of the costs of nuclear accidents began after the Chernobyl accident in Ukraine in 1986. This accident, as with the accident at Three Mile Island in 1979, made people around the world aware of the fact that a nuclear accident was possible on reactors used for electricity production. Subsequently, a number of calculation codes were created (MACCS then MACCS2 in the United States, Cosyma and later Pace in the United Kingdom and Rodos in Germany, Oscaar in Japan, C3X in France, etc.). The objective of these is to assess the localization, extent and level of radiological contamination found in contaminated areas after a nuclear accident, as well as the degree to which the contamination has been transferred to the environment, food products and humans.

After the Chernobyl accident, the authorities decided to implement actions seeking to protect the environment and the population. For example, during the radiological emergency phase, it was necessary to quickly evacuate the people living near the plant that had suffered the accident to keep them from being exposed to the plume of radiation, by then in the post-accident phase, to move the populations living in the areas most contaminated by the accident farther away in the long-term, to decontaminate urban and agricultural areas, etc. The actions taken to remediate the contamination come at a cost, but they make it possible to reduce the effects of the accident on the environment and people, therefore ultimately reducing the costs

Nuclear Economy 2,
coordinated by Jacques PERCEBOIS and Nicolas THIOLLIÈRE.

of accidents (such as costs of relocating populations, costs of agricultural losses, etc.).

Therefore, following the logic of evaluating the costs and benefits of the contamination remediation actions, it would be necessary to balance the cost of implementing these actions with the costs that this remediation allows us to avoid.

There is no single cost for a nuclear accident, but rather multiple costs that depend on the objectives pursued and the context of the analysis. The estimation of the cost of a nuclear accident thus only makes sense if the objective of the evaluation is specifically given, the calculation assumptions are defined and the results are explained.

4.1. Fields of economic analysis

Each economic study is defined by the objectives it seeks to accomplish, with the costs to be taken into account in the analysis varying based on the objectives pursued in the cost study.

Thus, in a situation in which an accident has occurred, the objective of nuclear accident cost studies is to optimize the contamination remediation strategies. To accomplish this, cost calculation modules have been added to the codes for calculating human and environmental consequences (Chronc for MACCS2, Coco-2 for Pace, Economo for Rodos, Arpagon for C3X). Only the costs directly associated with the radioactive contamination of the territory are taken into account, so that they can be compared with the costs of implementing these strategies.

In addition, some economic studies seek to determine whether the existing insurance mechanisms can provide sufficient coverage for all the damages that can be expected in the event of an accident. In this perspective oriented toward compensation for victims, only the costs that can be considered as damages in the sense of nuclear civil liability are taken into account in the calculations. For this to be the case, there must be a direct causal link between the accident and the damages suffered.

Finally, other economic studies aim to compare different strategies for the power supply mixture or evaluate the amounts of expenses that can be made to prevent the accident and are part of the improvement of the safety of nuclear installations. In these two cases, the objective is to quantify all the externalities associated with electricity production. This is the widest scope of costs to be taken into account in the calculations.

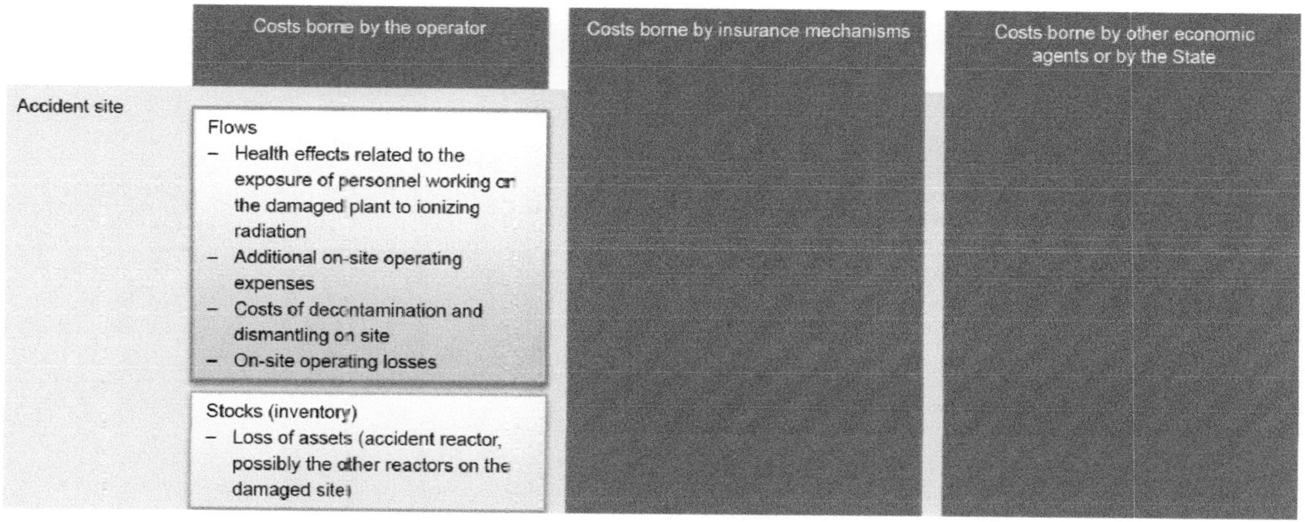
Costs borne by the operator
Costs borne by insurance mechanisms
Costs borne by other economic agents or by the State
Accident site
Flows
– Health effects related to the exposure of personnel working on the damaged plant to ionizing radiation
– Additional on-site operating expenses
– Costs of decontamination and dismantling on site
– On-site operating losses
Stocks (inventory)
– Loss of assets (accident reactor, possibly the other reactors on the damaged site)

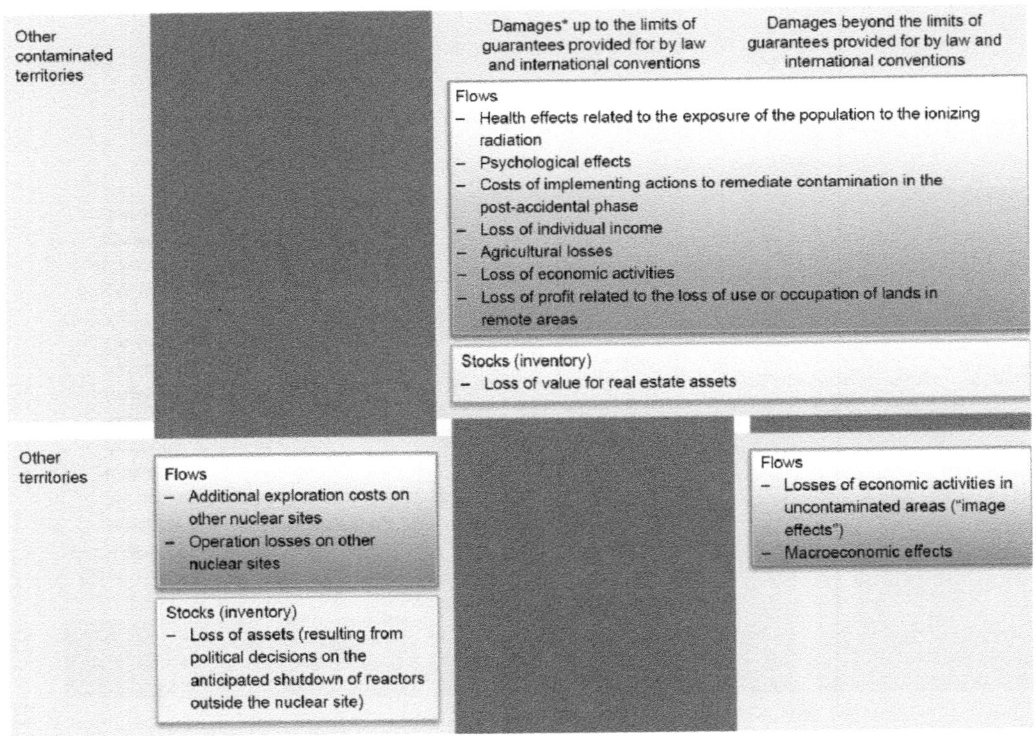

Figure 4.1. *The different economic costs caused by a nuclear accident. For a color version of this figure, see www.iste.co.uk/percebois/nuclear2.zip*

Figure 4.1 gives a schematic representation of the various economic consequences of a nuclear accident, which depend on the geographical area considered (the nuclear accident site, other contaminated areas and other uncontaminated areas) and on the economic agents who bear these costs (the nuclear operator, insurance mechanisms, other economic agents and the state).

Some costs can be referred to as "direct" costs, because they are directly associated with the radiological contamination of the territory. This is particularly the case with damages. In general, these costs can be compensated by the nuclear civil liability insurance mechanisms, up to a guarantee ceiling (APREF 2015).

Conversely, other costs can be described as "indirect" costs. Thus, in the agribusiness and tourism sectors, "image costs" will depend on the decisions by consumers and distributors to carry out boycotts for fear of radioactivity, which may include territories completely free of contamination. In the electricity production sector, the operator of the plant suffering the accident may decide to temporarily or permanently stop operating the damaged reactor, or even the other reactors on the site, while political decisions concerning the national electricity mix may be taken by the authorities after the accident. Finally, the macroeconomic effects are rarely attributable to the nuclear accident alone, because they are part of a broader economic environment, that of the country at the time of the accident.

4.2. Uses of nuclear accident cost assessments

4.2.1. *Optimization of post-accident management strategies*

The optimization of post-accident management strategies is the historical use of economic evaluations of nuclear accident costs. Software used for calculating the consequences of accidents, such as MACCS2, Pace, Rodos, C3X, etc., which calculate the atmospheric dispersion of radioactive elements in the atmosphere, how they are deposited on the ground, their transfer to the various compartments of the biosphere, and their subsequent effects on humans, have been supplemented with a module for estimating the off-site economic costs directly associated with the radiological contamination of the territory.

The objective is to evaluate and compare the different scenarios for the remediation of the contamination, and to compare them to a reference scenario where no remediation action is taken. These economic studies thus provide, on the one hand, the costs of implementing remediation scenarios, as well as the costs of transport, volume reduction and waste storage. On the other hand, they evaluate the effect of these remediation scenarios on the consequences following the accident, and thus provide an estimate of the costs avoided through the implementation of these remediation strategies (in terms of relocation of populations, agricultural losses, losses of economic activity of companies, health effects, etc.). This information serves as input data for cost–benefit studies intended first to determine the economically feasible remediation strategies (those whose net cost of benefits is negative or at the limit equal to zero) from all such strategies, then among the economically feasible strategies, those that are the most economically effective (those whose cost/benefit ratio is the lowest).

In this approach, only the costs directly associated with the contamination of the territory and outside the perimeter of the damaged installation are taken into account. The other costs (costs relating to the damaged installation and costs that do not directly depend on the radiological contamination of the territory) are not modified by the implementation of remediation strategies. Therefore, they are not taken into account at the level of the cost estimates carried out to evaluate and compare remediation strategies.

4.2.2. *Insurance mechanisms for covering a nuclear accident*

4.2.2.1. *Nuclear civil liability*

With the development of civil nuclear energy, a particular legal framework has been created to encourage the industry to develop this technology, as well as to offer immediately available funds for victims to be compensated quickly in the event of a nuclear accident, without exposing it to expenses large enough to potentially jeopardize its operations. The regime involved here is the specific regime of nuclear civil liability (nuclear RC).

The specific civil liability regime derogates certain points of the common law of civil liability. In this way, it defines the legal entity that must compensate the victims in the event of a nuclear accident and the competent court, the amounts available for this compensation and the period during which the victims have the right to claim compensation for damages. The creation of this specific regime resulted from the observation that the rules of the common law of civil liability could hinder rather than help victims of nuclear damage to obtain appropriate compensation in a timely manner. Common law is effectively based on wrongful acts; it implies that perpetrator of a harmful act is required to provide full compensation for the damages caused, and the requirement for the party seeking compensation to provide proof of the existence of a general fact attributable to the party at fault. Applied to nuclear power, these principles lead to insufficient insurance capacities to cover all compensation needs while posing solvency issues for insurers; potentially devastating losses for companies found responsible for the accident; and long and expensive judicial procedures to determine the parties responsible for the accident. Ultimately, the victims would receive less compensation with greater delays in the event of an accident. When a country does not adhere to any nuclear civil liability regime, the law that applies is civil liability, with all its limits.

In France, nuclear civil liability is governed by the Paris Convention (1960), the Brussels Convention (1963) and the Environmental Code. These conventions seek to create a global nuclear civil liability regime to address the complexities created by the potentially cross-border scope across which nuclear damages may be caused. They make it possible to harmonize legislation on civil nuclear liability, and thus promote equal treatment for victims and operators around the world. They define the competent jurisdiction in the event of an accident and ensure that legal rulings are carried out. They make it possible to increase the funds available for the compensation of victims through contributions to an international fund. These conventions are characterized by five main principles as follows:

– The nuclear operator is exclusively responsible for the damages, which means that all liability for all damages suffered by third parties as a result of an accident lies with the operator, who is the solely responsible for such damages.

– The operator's objective liability exempts the victims from having to prove that the operator has committed a fault, with the operator being automatically recognized as liable for the accident; it is only a causal link between the accident and the damage suffered that must be proven.

– The amount of the guarantee provided by the operator must be higher than a minimum value set by law and/or international conventions; in most cases, the operator's liability is limited (capped) in its amount and with regard to time.

– The operator is required to guarantee the financing of its civil liability, such as by setting up a fund.

– The jurisdiction is unique, which is to say that the courts with jurisdiction are those of the country where the accident took place.

For countries that have adhered to the Paris Convention and the Brussels Convention, the compensation system is made up of three segments of compensation. The first tranche is guaranteed by the nuclear operator up to a compensation cap; in France, this cap is 91.5 million euros (22.5 million euros for reduced-risk installations and transport). The second segment is guaranteed by the government of the country where the accident took place; in France, this segment of compensation is a minimum amount of roughly 115 million euros. Finally, the third segment is guaranteed by the signatory countries of the Paris and Brussels Conventions, up to a maximum amount of roughly 150 million euros. Thus, the maximum amount guaranteed under this system is set at around 350 million euros. Given the potentially grave nature of a nuclear accident, the government may have to play the role of final reinsurer, as it does for natural disasters. It would then bear, in addition to the amounts of the second segment, all the compensation owed to the victims beyond the caps provided for by international conventions.

No insurance company has the financial capacity to bear nuclear risks alone. The insurance providers therefore group themselves into *pools* at the national level (in France, this pool is Assuratome), which in turn use the capacities of their foreign counterparts to cover all the amounts necessary for operators. The insurance company can also transfer part of the risk to a reinsurance company. In France, EDF uses a captive reinsurance company (a subsidiary of the EDF group) to cover two-thirds (60 million euros) of the amount of its civil liability (91.5 million euros). For the remaining one-third (31.5 million euros), it uses the insurance market (the Assuratome pool of insurers) and the European mutual group ELINI.

The nuclear civil liability regimes cover damages caused to persons and those caused to private property outside the damaged facility up to the guarantee cap provided for by law and by international conventions, and within a limitation period of 10 years from the date of the incident. Recent international conventions (the protocols of 2004) extend the limitation period for personal injury to 30 years and provide for a wider coverage of damages in order to guarantee that reparations are provided for as many types of damages are as possible. First, personal injury is compensated, within the limits of the established compensation caps. Second, damages to property are compensated – if there are still funds available after compensation for personal injury, and within the limits of the established compensation caps.

Excluded from the scope of compensation are damages suffered at the site of the nuclear facility, all consequences that cannot be directly, causally linked to the accident, and all damages resulting from a nuclear accident caused by an act of war. Material damages, breakage of machinery, interruption of activity, etc., on the accident site are not covered by nuclear civil liability. The operator is obliged to assume such liabilities, such as by taking out additional insurance policies including "machine breakage", "operating losses" or "construction risks". Service companies involved in the construction or maintenance of a nuclear installation may be held liable for damages to goods occurring on the site.

4.2.2.2. *Compensation for damages from the Fukushima accident in Japan*

With regard to the legislative framework for compensating damages, Japan is one of the countries that has not entered into the nuclear civil liability or the corresponding international conventions, but it nevertheless has two major laws that regulate compensation for damages caused by nuclear accidents: the compensation law of 1961 and the law on the agreement of indemnities. The main difference between the system applied in Japan and nuclear civil liability stems from the fact that the Japanese operator is considered to have unlimited financial responsibility in the event of a nuclear accident. For each power plant in the network, TEPCO must ensure the availability of funds in the amount of 120 billion yen (or about 1 billion euros). The limit period is set at 20 years after the accident, and legal action can only be filed within 3 years after the victim has suffered damages and has become aware of such damages and the person responsible for them. It should also be noted that a clause is provided that exempts Japanese operators in the event of "a serious natural disaster of an exceptional nature and social revolt". In the case of the Fukushima accident, this liability exemption clause was not applied by the government but was invoked by the insurance company. Thus, it was the state that had to replace the insurer in providing the payment of the 120 billion yen as financial guarantee (Nomura 2014).

In the case of a large-scale accident such as the accident at Fukushima, the funds secured by the operator turned out to be very much insufficient for supporting the costs. Also, a system of "subsidies for aid" was implemented: a fund for "assistance for liability for nuclear damage" was created. To this end, two laws were passed in 2011 in order to complement the 1961 law: the "law on the creation of an aid organization for compensation for nuclear damage", and the "law on emergency compensation for victims of the 2011 nuclear accident". The law establishing an aid organization defines the general framework for financial assistance, allowing TEPCO to respond to victims' compensation claims, while the emergency law stipulates the mechanisms for rapid payment by the State of provisional compensation to individuals who are victims of the nuclear accident.

The law establishing an aid organization specifies the conditions for the assistance offered by the state to the operator of a nuclear power plant in the event that the total amount of compensation required exceeds 120 billion yen. The entity created by this bill is jointly funded by the government and the Japanese electric companies, including TEPCO. Its objective is to provide the necessary financial assistance to a nuclear operator which is required to compensate the victims of nuclear damages that have been suffered. In this way, the government can provide government bonds[1] to the aid organization so that it can raise funds. This assistance can take several forms, such as the provision of liquidity, buybacks of company stock, loans, the acquisition of company bonds and loan guarantees. In practice, providing liquidity has been the preferred solution for TEPCO. By providing funds to the organization in this way, the State indirectly replaces TEPCO, assuming partial responsibility for the compensation for damages and allowing TEPCO to continue its activities. To have access to such funds, TEPCO commits to paying exceptional contributions, with amounts that are much higher than ordinary annual contributions. These can be compared to reimbursements of the financial assistance that is received. This mechanism leaves TEPCO with the responsibility of compensating the victims in accordance with the legislation previously in force, but prevents the company from entering into liquidation, since it is now able to benefit from the financial support of the entity that has been created while also continuing to produce electricity. If it is chosen for the company to enter into liquidation proceedings, the repayments of the bondholders (the holders of TEPCO's company capital) will be given precedence, giving rise to the risk that the victims will be insufficiently compensated. The creation of an aid organization ensures more effective compensation for victims, provided over a longer period (CCI France Japan 2011).

The law on emergency compensation for victims of the 2011 nuclear accident allows the state to pay interim compensation to people who have suffered damages caused by the nuclear accident at the Fukushima power plant in order to ensure that compensation is provided as soon as possible. This emergency compensation amounts to more than half of the amount assumed that the victims would be entitled to receive. These provisional allowances are payments made by the State on behalf of TEPCO, from which it will then be entitled to claim reimbursement.

With regard to the methods for claiming compensation, three ways for claiming compensation exist in the case of the Fukushima accident. The first path provides for direct negotiations between the victim and TEPCO, as provided in the 1961 law. Victims must complete a claim file for compensation from their pre-trial judge and attach the supporting documents. Once this file has been studied, TEPCO offers compensation to the victim based on the guidelines produced by the Dispute

1 Government bonds are debt instruments issued by a state to finance its expenses by borrowing funds from the financial markets. In theory, government bonds are considered safe investments, because the probability of a state going bankrupt is marginal.

Resolution Committee (CRD)[2]. The majority of the victims are compensated using this method of compensation: 2.4 million claims were recorded in this way by TEPCO between 2011 and 2015. The second path goes through the CRD, which addresses cases that do not fall within the legal framework of the 1961 law, cases where the victim has refused TEPCO's offer and cases where the victim does not want to work with TEPCO. This system is made up of 200 lawyers who mediate between TEPCO and the victims, and junior lawyers who are given the role of investigating the files. These mediators provide a level of compensation based on both the damage compensation guidelines an on a general standard, a document complementing the classic compensation guidelines. For example, the system proposed by the CRD makes it possible to award compensation for moral damages to people who have voluntarily left a territory (and are therefore not considered forced evacuees) or to increase this compensation for people falling within particular profile (such as pregnant women). A total of 15,831 applications were filed between 2011 and April 2015 using this compensation system (12,809 cases closed in 2015). It should be noted that the dispute resolution committee also has two other tasks: the drafting of guidelines intended to define the overall nature of the damages and the assessment of the damages (Ishii and Morlans 2013). Finally, the third and final path is through the legal system. It is taken by people who are dissatisfied with the compensation offered by TEPCO, who consider the compensation offered in the second path too low or who do not wish to make their claim through the two previous paths. Taking legal action also allows compensation for non-material damages such as suicide, for which it is difficult to prove a direct causal link between the accident and the event, which is required by TEPCO to grant compensation. Rheuben and Nottage (2013) also indicate that the option of going to court is used by people who live outside Fukushima Prefecture or the city of Tokyo and who cannot easily access the dispute resolution committee.

With regard to the scope of compensation coverage, the CRD has produced guidelines to determine the extent of the damages attributable to the accident and subject to compensation. The first version was issued on April 28, 2011. The document has since been updated based on the situations that have been examined. The latest version of which we are aware is the version of December 2013. These guidelines present six types of compensable damages.

2 After the Fukushima accident, the Japanese government implemented several different mechanisms for improving how the system for providing reparations for nuclear damages functions. The Dispute Resolution Committee (CRD) was established in April 2011 to compensate for nuclear damages, and it has published several guidelines on the extent of damages subject to reparations. The Dispute Resolution Committee was established in September 2011 and has already resolved more than nine thousand cases through mediation.

4.2.2.2.1. Damages to life

For damages related to life in the territories affected by the accident, the guidelines list six types of zones: (1) the "evacuation zone", including the area within a radius of 20 km around the Fukushima Daiichi power plant where people were evacuated and the 10 km area around the Fukushima Daini power plant (reduced to an 8 km radius around the power plant on April 21, 2011); (2) the "sheltering zone", the area between the 20 km and 30 km radii around the Fukushima Daiichi power plant, where residents were instructed to take shelter in their residences. Taking into account the difficulty of living in an area and having to comply with such instructions, these territories were converted to a deliberate evacuation zone and an evacuation preparation zone as of April 22, 2011; (3) the "deliberate evacuation zone" beyond the 20 km radius around the Fukushima Daiichi plant, where there is a risk that the cumulative radiation levels will exceed 20 mSv during the year following the date of the accident leading to the delayed evacuation of the inhabitants; (4) the "evacuation preparation zone", located between the 20 km and 30 km radii from the Fukushima Daiichi power plant, where people must be ready to be moved at any time in case of emergency; (5) the "recommended evacuation points" are "hot spots" where the government declared a state of alert and requested the voluntary evacuation of residents; finally, (6) areas where the local government has asked residents to evacuate temporarily.

In these areas, compensable damages include costs directly related to the public authorities' evacuation decisions and losses of economic activity. The costs directly linked to the evacuation decisions of the public authorities are the expenses for medical examinations intended to assess the exposure of the populations (the examination itself, transport, etc.); the expenses related to the evacuation (transport, relocation accommodations, increases in the cost of living, etc.); the expenses related to temporary access to the access restriction area (transport, accommodation for the nights before and after the visit); the expenses related to the return to the accommodation; injuries or deaths resulting from the evacuation and psychological damage. The losses of economic activity include decreases in income (value added); the extraordinary costs related to a disruption of activity or to changes made to respond to the situation; the costs incurred after the instructions given by the government were lifted can also be compensated; the damages related to the inability to work (salary reduction), the expenses generated by the tests carried out to control the contamination of the property; and decrease in property values.

Among all these damages, psychological damage is the category where the compensation amounts are most clearly defined. Such compensation is paid to victims without them needing to present proof of psychological disorders. Thus, for people residing in contaminated territories at levels below 5 mSv, the compensation is JPY 100,000 per person per month for the first 6 months after the accident, or JPY 120,000 if the person is accommodated in an evacuation center (in a gym, for example), and JPY 50,000 per person per month thereafter. These allowances are collected for a "reasonable" period after the date when the ban on access to the territory is lifted. This "reasonable" period is set at 1 year in the latest supplemental version of the guidelines, published in December 2013. As an alternative, people evacuated from areas where maintaining residence is prohibited (with cumulative radiation levels between 20 mSv and 50 mSv) can choose to receive a flat rate of 2.4 million yen (approximately €20,000) for 2 years (Nomura 2014). Finally, for people evacuated from areas where it is difficult to return one day (cumulative levels of irradiation > 50 mSv), the compensation totals 10 million yen (about €80,000) per person.

4.2.2.2.2. Damages arising from the creation of exclusion zones

The damages related to the creation of marine and air exclusion zones (circular zones with a radius of 30 km around the Fukushima Daiichi power plant) include the decline in economic activity and the damages related to the inability to work.

4.2.2.2.3. Damages related to restrictions on the marketing of agricultural, forestry and fishing products

The damages related to restrictions on the marketing of agricultural, forestry and fishing products consist of (1) economic losses, whether direct (cessation of an activity or loss of income concerning goods already produced at the time of the ban) or indirect (recalls of products, waste management, changes to production processes, the supply of raw materials). If costs are incurred after the lifting of the restrictions but are related to them (such as decontamination), they are also compensable; (2) damages related to the inability to work; and finally, expenses related to tests carried out to control the contamination of property (products, cargo, containers, etc.).

4.2.2.2.4. Damages related to effects to image, negative reputation and hearsay

The damages related to effects to image, bad reputation and hearsay include (1) image effects related to the fishing, agriculture, forestry and food production sectors, as well as the costs of radioactivity tests to products; (2) image effects that are the responsibility of tourism for the prefectures of Fukushima, Ibaraki, Tochigi and Gunma; and (3) image effects affecting industry and services; export image effects.

4.2.2.2.5. "Indirect" damages

The "indirect" damages represent the losses suffered by indirect victims, who have maintained economic relations with direct victims. The damages taken into account in this case are income losses and damages related to the inability to work.

4.2.2.2.6. Damages resulting from people's exposure to radiation

The damages resulting from people's exposure to radiation include all costs related to access to care for people affected by a radiological risk.

The CRD guide provides that for people or companies who have suffered damages but also have earned profits as a result of the accident, the profits must be deducted from the amount of compensation given for damages.

It is also interesting to note that some types of damages not listed in the *Interim guidelines* may be compensated on a case-by-case basis if there is a causal link between the damage and the nuclear accident. It is then necessary for the victims to prove the causal link with the accident that is not linked to the earthquake and the tsunami.

Finally, with regard to the amount of compensation, at the macroeconomic level, the investigation committee on TEPCO's finances and management estimated in 2011 that compensation would amount to 4,500 billion yen for the first 2 years (i.e. about 34 billion euros) (OECD 2012b). This estimate led the Japanese government to, among other things, issue 5,000 billion yen of national bonds (about €40 billion) to provide for these future expenses. In this cost assessment, the compensation amounts were estimated at 9.5 billion euros for the first year, 7 billion euros for the second year and the following years, and 20 billion euros for other so-called "transitional" losses such as real estate losses (about 4.3 billion euros) as well as damages from hearsay in the agricultural, fish farming and forestry fields (estimated at about 10 billion euros).

In 2012, Kenichi Oshima (2012) presented an estimate of the costs from the Fukushima accident. Among the cost lines are the compensation paid to the victims for each year after the accident. These amount to 25 billion euros for compensation related to very short-term damages; 10 billion euros for compensation related to damage occurring in the first year; 9 billion euros for compensation related to damage occurring in the second year; and 13 billion euros for compensation related to damage occurring from the third to the fifth year. These estimates lead to a compensation cost of about 58 billion euros for the 5 years following the accident.

In October 2016, the federation of Japanese electric companies (FEPC) estimated the amount of compensation at about 80 billion euros (Hasegawa 2016).

The most recent study to our knowledge is that of the Japan Center for Economic Research (JCER) of March 2017, which carried out an estimate of the cost of the Fukushima nuclear accident and who considers that the compensation amounts to about 8,000 billion yen (or about €62 billion). It is interesting to note that this estimate is similar to the one that the Japanese government made in December 2016 (JCER 2017).

At the microeconomic level, a family (consisting of three or four people) receives 430,000 euros for being located in the "difficult to return" zone; 252,000 euros for the "restricted residence" and "preparation for restrictions to be lifted" zones; 54,000 euros for areas located between 20 km and 30 km from Fukushima; and 7,200 euros outside the evacuation zones. The report prepared by Greenpeace (2013) mentions that these amounts are paid both as compensation for moral damages and for losses of real estate property values. Some cities, such as Tokyo, provide assistance to evacuees (voluntary or compulsory), which consists of free housing as well as a monthly gratuity (which was required to continue until March 2017). Finally, evacuees can receive an allowance to visit their former residences in the exclusion zone. According to one of the people interviewed for the Greenpeace report, the compensation totals 118 euros per month.

4.2.3. *Evaluation of investments for the safety of nuclear installations*

Economic cost assessments of nuclear accidents can also be used to assign an economic value to the effects of the accident on society as a whole. This involves calculating the externalities associated with the production of nuclear electricity. An externality is the external effect that a productive activity generates for other economic agents (other than the producer), without there being any financial exchange between the producer and the other economic agents to compensate for this effect. Externalities can be positive or negative; an example of a negative externality is pollution. The nuclear accident is considered as a negative externality of the production of nuclear electricity, since most of its effects are borne by economic agents other than the operator: by the government, which plays the role of ultimate reinsurer in the event of a nuclear accident beyond the compensation ceilings provided for by the nuclear civil liability insurance; and by the general public.

To estimate the amount of possible investment expenses to be made for the prevention of the accident, it is not enough to simply evaluate the costs of nuclear accidents. It is also necessary to take into account the likelihood of an accident. By multiplying the accident costs by the likelihood of an accident, and adding together the values obtained in this way for all plausible accident scenarios arising from a nuclear facility, we obtain the mathematical expectation of loss related to the nuclear accident for this installation.

This measure of the risk formed by the mathematical expectation of loss is satisfactory for known, mutualizable, diversifiable, non-systemic and non-catastrophic risks according to the Arrow-Lind theorem, which concludes that the price of the risk (in other words, the risk premium according to insurance terminology) tends toward zero when the risk is perfectly diversifiable and mutualizable, and subject to the pressure of competition.

However, nuclear risk does not satisfy any of the conditions stipulated above: knowledge of it is imperfect, as is its ability to be diversified, and mutualized, it is systemic and it is potentially catastrophic in magnitude. Consequently, the risk premium applied to the risk of a nuclear accident is not zero. The literature in France on the modeling and evaluation of this risk premium includes most notably (Eeckhoudt et al. 2000; Bizet and Levêque 2017; Louass and Picard 2017). These models use utility functions to describe individuals' nuclear risk preferences. These preferences are modeled in the form of a HARA (*harmonic absolute risk aversion*) utility function by Louass and Picard (2017) and a CRRA (*constant relative risk aversion*) utility function by Eeckhoudt et al. (2000) and Bizet and Levêque (2017). These utility functions take into account the aversion of individuals and the public decision-maker toward nuclear risk. Thus, for example, by using values between 1 and 5 from (Szpiro 1986; Levy 1994; Louass and Picard 2017) for the nuclear risk aversion parameter, the contribution of the risk premium to the public cost of risk represents approximately 70–80%, which confirms the importance of taking into account risk aversion in calculating the cost of risk.

What does this risk aversion represent? It represents policymakers' aversion to inequalities in consumptions between individuals. Policymakers take into account not only the preferences of individuals, but also their own preferences. In particular, the existence of consequences of non-mutualizable nuclear accidents implies that it is not possible to tax individuals according to their exposure to risk (this would be experienced as a double penalty, because the people taxed the most will also be those who will be most affected in the event of an accident). Taxes are thus set in a homogeneous and centralized way, and taxation is used as a tool to rebalance inequalities in risk exposure. In other words, policymakers prefer to tax individuals with low exposure to nuclear risk more than what would be optimal, and to tax individuals with high exposure to this risk less than what would be optimal, even if the latter group of individuals is very small compared to the population as a whole.

4.3. Methods for estimating the costs of a nuclear accident

4.3.1. *General considerations*

The coupling of the assessment of potential physical and economic impacts to the frequencies of nuclear accidents is part of a comprehensive approach known as a "probabilistic safety study". After this is done, it is then possible to evaluate the costs of emergency countermeasures and environmental decontamination actions to be potentially taken by the authorities in the event of a nuclear accident, as well as their benefits in terms of reducing the consequences of the accident.

The assessment of the economic costs of a nuclear accident is then divided into three main stages. The first step consists of estimating the physical consequences directly associated with the radiological contamination of the territory, that is to say, the extent of the territory that has been contaminated and the number of people living in these territories according to their level of contamination, the quantity of agricultural commodities lost, the number of businesses and the number of residential units present in the contaminated areas or the committed dose for the population.

> The objective of probabilistic safety studies is to contribute to the assessment of the risks associated with nuclear power reactors for the surrounding populations and the environment. The method consists of comprehensively identifying and analyzing the possible accident scenarios, estimating their frequency of occurrence (often by grouping them by families), and evaluating the related consequences inside the facility (the fusion of the core) or outside of it (releases of radioactivity into the environment). The level 1 probabilistic studies are concerned with the consequences in terms of fuel fusion; the level 2 probabilistic studies are concerned with radioactive discharges outside the installation; and the level 3 probabilistic studies evaluate their consequences for the surrounding populations, the environment and the associated costs.

Box 4.1. *Objectives and methods of probabilistic safety studies*

The second step is to monetize these physical consequences using the appropriate methods. One or more monetization methods are adapted to each consequence. The use of direct methods is always preferable to indirect methods when possible, as they do not induce a bias in the evaluation. Direct methods are based on exchanges carried out on the markets, making it possible to set prices that can be compared to directly estimate the monetary values useful for cost analysis. In some cases, there is no direct method to assess the monetary value of a good or a service rendered by the environment. Sometimes direct methods do not capture the full economic value of a good or service. This is particularly the case for non-use values that exist independently of the direct use of a good or a service, such as the desire to leave in place certain environmental resources considered essential to future generations for their benefit. These quantities are of a non-market nature – they have no market price – which makes it impossible to estimate their value directly. It is therefore necessary to use indirect methods, with the objective of quantifying the gain or loss of well-being by an individual resulting from an improvement or deterioration in the quality of a service provided by a natural resource in monetary terms. These indirect methods are based on the theory of consumer behavior (Mas-Colell et al. 1995) and seek to identify the value of a good for individuals from the satisfaction – the usefulness – that it provides them, so as to better take into account environmental issues in public decisions and private choices. From among all the existing methods, some evaluate economic flows, the value of monetary exchanges between economic agents over a given period of time; the flow is then renewed from period to period. Other methods evaluate inventories, namely the valuation at a given date of the goods owned by an economic agent; these inventories are thus likened to capital. The value of an economic good can therefore often be estimated either as a cash flow or as an inventory. It is recommended not to add up the flows and the stocks, in order to avoid double accounts.

Finally, the third and final stage seeks to evaluate the other accident consequences, which do not depend on radiological contamination, but which result from political decisions involving the electrical mix and its possible evolutions following the accident, as well as consumer preferences in terms of agricultural product and tourist consumption in particular. These latter effects are often referred to as "image effects", because they result from changes in the food products consumed or the tourist regions frequented, caused by the fear of radioactive contamination. This third step is not systematic: it is carried out only when the economic study seeks to evaluate the externalities related to electricity production.

4.3.2. *Costs at the accident site*

From among all the consequences of a nuclear accident, the costs incurred at the site exceed those contained within a geographical area owned by the operator. They include material damages (the breakage of machinery), the costs of cleaning the site and the additional expenses of dismantling the damaged reactor, or even the other reactors on the site due to radiological contamination, the real operating costs (maintenance, monitoring) of the damaged site and the health costs for personnel exposed to radiation.

4.3.2.1. *Material damages*

They represent a shutdown of the means of electricity production on the accident site, temporarily or permanently, after a nuclear accident. They can be calculated as a loss of capital: this is therefore a kind of inventory that represents the economic value of the reactor(s) at the time of the accident. This inventory can be estimated as the updated sum of the investment amounts devoted to the construction, maintenance and safety of the damaged reactor, from which we subtract the share of these amounts that has already been amortized during the years of operation of the reactor. Another method is to estimate this stock as the replacement value if it should be necessary to replace the damaged reactor with a new reactor. These production losses can also be estimated by the method of replacement cost. This method evaluates the difference in the price of MWh which can no longer be produced by the damaged nuclear reactor, and which must be replaced at a higher cost by another means of production. In this case, the economic losses correspond to a flow based on the real costs on the electricity market, or on the reference costs of electricity production (OECD 2015). They are calculated over the entire period from the date of the accident to the date of the initially planned end of operation of the reactor in the event of its final shutdown, or over all the years during which the reactor is temporarily shut down. They are then updated to the date of the accident, and then added together.

4.3.2.2. *Cleaning costs*

The on-site cleaning costs are the costs of the decontamination operations on the accident site. They are grouped into several categories: the emptying of the reactor building, the decontamination of the buildings, the unloading of fuel, the treatment of contaminated water and the possible construction of a sarcophagus. They are estimated as the sum of the flows of expenses devoted to the various cleaning operations as of the date of the accident.

4.3.2.3. *Additional dismantling costs*

The additional costs of dismantling the damaged reactor, or even the other reactors on the site, represent the additional costs of dismantling due to the radiological contamination of the site (the dismantling operations are slower, and the personnel involved must be protected from exposure to ionizing radiation). These additional expenses are then updated to the date of the accident, and then added together. In addition to this, as part of the immediate decommissioning strategies, the differences in the updates are caused by expenses that are incurred earlier than initially planned at the end of operation of the reactor.

4.3.2.4. *Residual operating costs*

They represent the amounts of expenses for maintenance operations and monitoring of radiological contamination on the site during the entire period of post-operation of the site, until the dismantling of the reactors. More broadly, they represent the additional personnel expenses generated by the changes to the management of the site. These cash flows are then updated to the date of the accident, and then added together.

4.3.2.5. *Health-related costs*

To assess the health-related costs, it is first necessary to estimate the number of health effects expected as a result of the accident. According to the classification of the publication No. 103 of the ICRP (2007), the effects of exposure to ionizing radiation may have certain effects after the exposure of individuals above a certain dosage threshold (deterministic effects or tissue reactions), several years after and at random (stochastic effects) or may also manifest themselves in the descendants of exposed individuals (hereditary stochastic effects).

The models used to estimate the number of deterministic effects are survival models (Rodriguez 2007) based on the estimation of the chance rate (Evans et al. 1985). These models are relevant for calculating the number of deterministic effects since they make it possible to estimate the frequency of the health effect that is feared to occur depending on the dose received and the characteristics of the exposed population. The functions used to model the number of deterministic effects in the exposed population are formulated from statistical models of empirical analysis of biological data. These data come from studies of the effects of medical exposure to ionizing radiation, epidemiological studies of survivors of the bombings of Hiroshima and Nagasaki, and radio-biological studies of animals. These data are processed statistically using dose–response functions; several types of dose–response functions are offered in the literature (logistic, Gaussian, gamma, etc.) (Wells 1976; Jones 1981). The results of the deterministic effects models are interpreted as probabilities because they reflect the different degrees of sensitivity of individuals exposed to ionizing radiation.

The models used to estimate the number of stochastic effects, that is, the increased incidence of fatal and non-fatal cancers in the population exposed to ionizing radiation, are linear models with or without a dose threshold below which no increase in the number of cancers is expected in the exposed population. According to an analysis of the NCRP, the LNT hypothesis[3] for the dose–response function in the low-dose spectrum for modeling stochastic effects is the most *pragmatic or prudent for the objectives of radiation protection* (NCRP 2018).

Once the numbers of deterministic and stochastic effects have been calculated, it is possible to monetize them using an appropriate method. Several methods exist to monitor the effects of ionizing radiation on human health, depending on the costs to be assessed. This involves first the estimating of the treatment costs: this is the market part of the health cost. To estimate these costs, it is possible to use market data, such as the prices of care, drug reimbursements or the costs of medical equipment. The costs of multiple different medical procedures included in the overall treatment of the patient can be added up to estimate the total cost of treatment of a health effect. The annual treatment costs are recorded and updated according to the duration of the treatment and the latency time between exposure to ionizing radiation and the appearance of the health effects.

The treatment costs are then added together with the opportunity costs, in other words the net losses for the economy created by the health effects. In the literature on the evaluation of the value of statistical life (Drèze 1962; Mariotto et al. 2011), the historical method for evaluating opportunity costs is through the use of "human capital". This method counts the productive losses of individuals affected by a health problem. For health effects that are fatal, the human capital method seeks to calculate the sum of all the wealth that cannot be produced by an individual between the date of that individual's death and the average life expectancy of an identical individual who was not exposed to ionizing radiation. The wealth produced is then represented by the net production, assessed as the difference between the loss of wages and the consumption of the deceased individual. In this method, the value that individuals bring to the economy (their marginal productivity) is assumed to be equal to their salaries. For non-fatal health effects, the calculation of the loss of production assumes that the individual consumes the same amount of goods every year during his/her entire illness, but loses part of his/her annual salary (in general, the proportion of lost salary is taken as equal to 50% of the total salary of this individual).

The "human capital" approach does not take into account intangible costs such as pain, grief and distress associated with health effects, and does not give a value

3 Linear model with no threshold.

for individual preferences in particular with regard to individuals' aversion to risk. To estimate these intangible components of the cost of health effects, it is necessary to use other approaches, known as "consent to pay" in exchange for a reduction in a health risk (such the risk of dying), or "consent to receive" in exchange for an acceptance of this risk. There are several methods to measure the values of consent to pay or consent to receive. These measure the preferences of individuals based on the observations of behaviors in real or hypothetical markets. The most obvious way to determine the preferences of individuals is to ask them questions directly, thus making their preferences clear. They can also be revealed through surveys; individuals are thus placed within the framework of hypothetical markets, in which they must express their consent to pay (respectively, to receive) for improvements (respectively, degradations) in their health. Contingent evaluations are then carried out[4] or events with discrete choices[5]. In some cases, it is also possible to determine the value of a health asset by studying real markets, when market assets are complements or substitutes for intangible goods to be valued. In practice, there are many biases that are associated with these estimation methods, such as hypothetical biases (the respondents may not believe the hypothetical situation proposed to them), anchoring biases (on stated values) and information and inclusion biases, etc.

The value of a statistical life (VSL), the value of a statistical life year (VSLY) and the value of a life year (VOLY) corresponding to the VSLY, adjusted according to the age of the individual at the time of diagnosis of the disease, are the three main

4 In contingent assessments, the amount that respondents would be willing to pay is asked explicitly, or different possible amounts are suggested and they are asked to choose the one they prefer. The questions asked are along the lines of: "How much would you be willing to pay every month to be able to go for a walk for more than an hour more than a kilometer away from your home?"

5 The discrete choice evaluation method is derived from the contingent evaluation method. It breaks down the asset into several attributes (time outdoors, distance from home, price to pay). Several scenarios are presented to the respondents, with different values for the attributes. The participants must then choose the one they prefer from among all the scenarios, or at a minimum, the one they hate the least. For example:

– Reference trend scenario (time outside = 8 h, distance from home = 50 km, death = 10,000/d, price to be paid = 0);

– Scenario 1: strict confinement (time outside = 1 h, distance from home = 1 km, death = 0, price to be paid = 500 euros);

– Scenario 2: Managed confinement (time outside = 1 h, distance from home = 1 km, death = 100/d, price to be paid = €200);

– Scenario 3: Christmas holidays (time outside = 3 h, distance from home = 20 km, death = 1,000/d, price to be paid = €100).

Any resemblance to the Covid-19 crisis in 2020 is purely coincidental.

types of values used to monetize morbidity and mortality in the world. These values are calculated by country using the consent to pay or consent to receive methods. According to Viscusi, the values of the VSL for the United States vary from 4 to 10 million dollars, with an average value of 7 million dollars (Viscusi 2005). The study (ExternE 2005) reports values of VSL between 1 and 5 million euros for Europe and North America, and recommends the use of a central value of 1 million euros for the VSL and 75,000 euros for the VSLY. An OECD study (2012) reports a range of 1.5–4.5 million dollars with a central value of 3 million dollars for the VSL in the OECD countries, and a central value of 40,000 euros for the VSLY in European countries.

4.3.3. *Off-site costs directly related to contamination*

4.3.3.1. *Agricultural losses*

In certain geographical areas affected by radioactive deposits, agricultural products are considered to be too contaminated with regard to threshold values set by the Euratom Directive of the European Community (Euratom 2013), which are themselves based on the principles set out by the International Commission on Radiological Protection (ICRP). In these geographical areas, the placing on the market and the consumption of locally produced food products are temporarily prohibited during the implementation of discharge controls (IRSN 2012). These discharge controls seek to ensure that the contamination of agricultural products intended for human or animal food is below the standards in force for marketable goods, and if necessary, to maintain restrictions on commercial sale wherever these standards are exceeded. These controls are preferably organized by production sector, and they can gradually lead to releasing certain sectors from prohibitions on sales.

The estimation of agricultural losses is carried out in several stages. Once the radioactive deposits on the ground have been calculated using software such as MACCS2, Pace, C3X, and so on, it is then possible to estimate the proportion of the contamination that will pass from the soil to agricultural crops through the leaves and roots, and then that which will be absorbed by the animals grazing in the pastures or which are used as animal feed. To this end, the data on radioactive deposits in the ground are combined with information on agricultural occupation of the territory to estimate the concentration of radioactivity of agricultural products at any point within the contaminated areas. These concentrations are then compared with the threshold values allowing for the marketing of agricultural products. Products contaminated beyond these threshold values are considered as losses; their monetary value is updated as of the date of the accident and then added up.

4.3.3.2. *Direct losses of business activities*

The direct losses of the activities of companies located in contaminated territories are defined as the sum of the cash flows (the added value) that are exchanged there and whose value could be totally or partially lost in the event of an accident. Three categories of territories can be determined based on their level of contamination.

Highly contaminated territories are those where the accumulation of dosages likely to be received over several years will lead to the relocation of the population outside these territories. With a few exceptions (management of the accident site, work done to reduce contamination, etc.), all economic activity will cease in these heavily contaminated territories. In moderately contaminated territories, the population can stay within the territories, but radiological checks are carried out before the goods produced there are put on the market. Some products cannot be marketed or exported because they are considered to be too contaminated, and the production in volume and in price of the various sectors of economic activity decreases. The lightly contaminated territories, far enough away from the accident site, are generally spared from radioactive discharges. Economic activity may remain generally stable there, or it may benefit from a postponement effect to economic activity that can no longer be produced in moderately or heavily contaminated territories. The hedonic evaluation method, which consists of comparing the changes in volumes and prices before and after a nuclear accident, is particularly well suited to estimating these losses. The loss is thus equal to the difference between the situation that actually occurred after the accident and the counterfactual situation, that is to say, the situation in which the accident had not taken place.

4.3.3.3. *Losses of real estate value*

The losses in real estate value of buildings located in contaminated territories are defined as the decrease in the value of real estate properties on a given date. As long as the real estate is not sold, no losses of economic value have been incurred. They only occur once, on the day of the sale. These capital costs are very different in nature from the estimated losses in the form of economic flows and they should not be added together. The hedonic evaluation method is particularly well suited to the estimation of these losses. The loss is thus equal to the difference between the value of the property that is actually observed at the time of sale, and the counterfactual value of this property, in other words, which would have been observed if the accident had not occurred.

4.3.3.4. *Health costs related to exposure to ionizing radiation*

These are the effects on the health of populations resulting from exposure to ionizing radiation. They concern both deterministic effects (even if almost no deterministic effects are expected in the general population) and stochastic effects (non-fatal cancers, fatal cancers and hereditary effects). The calculation method is exactly identical to that described above for workers exposed to ionizing radiation at the site of the power plant suffering the accident.

4.3.3.5. *Costs to remediate the contamination*

After a nuclear accident, the public authorities may decide to implement remediation strategies for contamination in the territories contaminated by the incident. For example, in urban areas this would involve washing pavements, walls and roofs, and removing the surface layer of dirt in parks and gardens, etc., in order to reduce the external exposure of the population to ionizing radiation as much as reasonably possible. In agricultural areas, for example, it is possible to perform deep plowing to reduce the absorption of radioactive contamination by crops and thus by animals. The objective is therefore to reduce the contamination of the environment, and ultimately to reduce the exposure of the population through the ingestion of contaminated food. These contamination remediation operations come at a cost, which must be estimated. In general, the unit data (euros/m^2 of contaminated surface) from the observations made of the accidents at Chernobyl in Ukraine and Fukushima in Japan provide the information needed to make the calculation. They are applied to the occupation of the territory present in the areas contaminated by the accident, where it is desired to evaluate the cost. It is also necessary to estimate the volume of waste generated by these contamination remediation actions, grouping them into at least two main categories: flammable waste and non-flammable waste. It is possible to reduce the volume of flammable waste before storage; this operation to reduce the volume of waste will concentrate the radioactivity. Therefore, at the end of the treatment, a small amount of highly radioactive waste will remain. For non-flammable waste, reducing the volume is more difficult; it will therefore be necessary to store a fairly large amount of waste, but this will be relatively uncontaminated. The cost of these operations to reduce the volume of waste, as well as the costs of transportation to storage sites and the costs of storing waste, must be accounted for.

Remediating the contamination offers benefits, which it is also necessary to account for in order to balance them against the estimate of the corresponding costs. For example, the decontamination of urban areas allows for populations that have been moved away to return more quickly; the decontamination of agricultural areas allows for a faster return to the market of agricultural products; and the health effects related to exposure to ionizing radiation are in turn reduced.

4.3.3.6. *Cost of emergency countermeasures*

The emergency countermeasures seek to reduce external and internal exposure by inhalation during the passage of the radioactive plume, beginning at the start of the discharges from the damaged installation and up until the deposit in the ground of all the radioactive elements rejected. These countermeasures are decided by the prefect during the phase of the threat of releases, when the players involved in crisis management actions predict an unfavorable progression of the state of the installation that may lead to radioactive releases into the atmosphere. These countermeasures are as follows: evacuation, sheltering and listening to the populations, and ingestion of stable iodine tablets to saturate the thyroid with non-radioactive iodine by populations located in areas potentially affected by the radioactive plume, depending on the dose levels that they would be likely to receive. Sheltering can also be implemented reflexively for certain situations where rapid releases into the environment would occur.

The costs of countermeasures remain generally quite low compared to the other costs related to the post-accident phase of a nuclear accident. To estimate them, it is sufficient to account for the expenses actually incurred in the event of an accident and related to the implementation of emergency countermeasures. These expenses do not take into account the prepositioning and renewal of the stocks of stable iodine at the regional level before the accident, which in France are the responsibility of the nuclear operator (EDF), and which are already included in the production price of nuclear electricity.

In the first hours and the first days after a nuclear accident on an electricity generation reactor, the most important contribution to the individual dose is that of iodine isotopes, which have short half-lives. Therefore, in general, the duration of the evacuation does not exceed a few weeks, after which the evacuated population is allowed to return home. However, in certain geographical areas, depending on the level of residual contamination, the authorities may initiate a long-term population removal program.

4.3.3.7. *Cost of the long-term relocation of populations*

As was the case in Japan after the Fukushima accident, the process of relocating populations takes place in three major stages that can span several years after the accident.

The first step is reception at emergency locations, such as gyms, events spaces and other public buildings. Part of the population could find refuge with family or with friends, or in a commercial hotel structure. In general, the conditions of reception at these spaces would be quite precarious. This type of accommodation is therefore intended to be replaced quickly. The costs of transport to the meeting places, the cost of accommodation and the cost of intervention teams are recorded at the level of the costs of emergency reception.

The second step is to build temporary accommodation, such as in prefabricated structures. These have the advantage of being able to be created fairly quickly, but they are not intended to remain permanently. The duration of temporary housing could vary between 2–5 years (WHO 2012) depending on the severity of the accident and the ability of the construction sector to respond to the increased demand for new construction. In addition to the cost of temporary accommodations, the cost of sewage, water and electricity supply networks for temporary housing must also be considered.

Finally, if there is no possibility to return, the temporary accommodations will become permanent relocation. The cost of this includes the cost of purchasing land, the construction of new housing and, if necessary, the cost of sewerage networks and the supply of water and electricity to these new housing units. In addition, there is the cost of increasing the capacity of public structures (particularly schools), hospitals, etc.

4.3.3.8. *Loss of the intangible value of environmental goods*

Environmental goods, such as forests, have a usage value that is a market value arising through the exchange of wood on the market. They also have an intangible value, which is related to both their use as a place of leisure (used for walks or picnics, for example) and a value of existence. The economic loss of a contaminated forest therefore goes beyond its simple market value. To assess the intangible value of environmental goods, it is necessary to make use of well-known methods in environmental economics, based on the declaration or revelation of individual preferences in terms of access, development of environmental goods, preservation of fauna and flora, etc. These are the methods of consent to pay or consent to receive, which have already been described above. They are intended to measure the value of an environmental asset for the surrounding population using the well-being it brings to them.

4.3.3.9. *Primary limitations*

The main limitation of all off-site cost calculations directly related to the contamination of the territory lies in the generally accepted assumption that both the population and the economic production are distributed homogeneously within an administrative entity (municipality, department, region). This hypothesis does not reflect the real geographical distribution of the population and economic production within this administrative entity. For example, within a district, certain areas are densely populated (the "towns") and therefore concentrate the housing stock, while other areas often located on the outskirts of the town are used either for agricultural production, or forestry, or for industrial production. The results obtained from this are therefore orders of magnitude.

4.3.4. *Other off-site costs*

4.3.4.1. *Image costs*

Image effects represent a type of consumer boycott of certain agricultural products or certain tourist destinations for fear of contagion. These effects concern marketable products (which are therefore below the threshold values for radiation used for the marketing of agricultural products) and territories located outside the evaluation and relocation zones. They are based on the greater or lesser degree of simplicity with which consumers are able to substitute certain products or destinations considered as potentially dangerous to their health with products from other regions or imported from other tourist destinations abroad. The further consumers are from the area affected by the radioactive discharges, the more they will tend to broadly associate the contaminated area with the entire region, or even the country where the accident took place.

To estimate these costs, two methods can be used. The first method consists of defining possible scenarios concerning changes in consumption after the accident, which describe the percentages of decrease in consumption and how they evolve over time. The use of several sufficiently contrasting scenarios makes it possible to obtain a range for estimating image costs. The difficulty with this method lies in establishing the parameters of these percentages of decrease in consumption. They can be estimated using past events (the accidents at Chernobyl and Fukushima), but due to the fortunate fact that the number of nuclear accidents is very low, this makes it difficult to infer or transpose consumer behavior from one country to another. These percentages can also be derived from other health crises (mad cow, Spanish cucumber, etc.), but again, these are not easy to transpose to the case of a nuclear accident. According to Bizet and Levêque (2017), it is appropriate to use the methods that have been developed in health economics and environmental economics to assign economic values to losses in well-being, that is to say, methods of declaring or revealing preferences.

4.3.4.2. *Political decision-making costs regarding the mixture of electric power supply*

After a nuclear accident, the public authorities may decide to modify the electric power mix, such as by reducing the share of nuclear power in the production of electricity. The electricity that can no longer be produced from nuclear reactors is replaced by other means of production, generally at a higher cost since the construction of new electricity production facilities requires capital costs. The calculation method is identical to that used to replace the electricity that can no longer be produced on the accident site, whether temporarily or permanently.

If no precise information exists on the possible changes to the electricity mixture after a nuclear accident, the best course of action is to identify several possible scenarios and then to evaluate the costs. The use of a limited number of sufficiently contrasting scenarios allows for a cost estimation range to be obtained.

4.3.4.3. *Indirect and macroeconomic costs*

The direct losses suffered by the various economic sectors after a nuclear accident create a shock to the local, regional or national economies, depending on the severity of the accident. This shock then ripples across the economy via the interactions between its different components. For example, farmers who can no longer obtain income from their agricultural lands will no longer buy seeds from their suppliers, who are possibly located outside the contaminated area; hotel owners will no longer use their cleaning services.

The initial shock as well as its ripple effects have an impact on employment, added value, public expenditure and revenues, and the balance of payments. To evaluate these, one possible solution is to use the input–output tables of the economy. However, the main limitation of this method is that it considers the structure of the economy to be static. This hypothesis has not been confirmed in the event of a serious nuclear accident. This is why the methods of general computable equilibrium may be preferred over the input–output method, but they make use of a large number of parameters that are not easy to calibrate.

4.3.4.4. *Psychological costs*

The psychological effects on health following a nuclear accident and/or the related countermeasures (evacuations, relocations) most notably include anxiety, depression and post-traumatic stress (Bromet et al. 2011). The magnitude of the psychological effects also depend on the perceived quality of the emergency response, the competence of the authorities, the feeling of helplessness while facing a poorly understood threat, etc. Psychological effects affect the victims' health, require medical treatment and can lead to significant direct and indirect costs (such as lost working days) for society. In addition to these costs, there are intangible costs, which represent the loss of quality of life and which can be measured in the form of QALY (quality adjusted life year). The value of a QALY varies between 0 and 1, where 0 represents death and 1 represents a healthy life. The loss of QALY is generally assessed using psychological questionnaires such as the EQ-5D, which can assess the impact of the state of health on the quality of life based on five key areas: mobility, personal autonomy, current activities, pain/discomfort and anxiety/depression.

4.3.4.5. *Primary limitations*

The main limitation of the estimation of off-site costs that are not directly associated with the radiological contamination of the territory is the calibration of the various parameters of the economic models. The observations made from the Chernobyl and Fukushima nuclear accidents may be used, but these two events alone do not provide enough information to infer values for accidents of different severity in different countries. That is why economists also use the feedback from events of a non-nuclear nature to infer values for the nuclear accident. Since this transposition is delicate, several scenarios are generally evaluated in order to provide an estimate of a range of possible costs.

The use of methods derived from the economics of health and the environment which seek to evaluate the preferences of the consumers, the population, following a nuclear accident, even if they are not without bias, constitutes an important advance toward finalizing these cost estimates.

4.4. Application to the Fukushima nuclear accident

4.4.1. *Synthesis*

The assessment of the costs of a nuclear accident is complicated in the case of the Fukushima incident by the presence of three interrelated events: the earthquake, the tsunami and the nuclear disaster. Therefore, some of the economic losses suffered by the Japanese economy due to these interrelated events are difficult to discern individually. This is particularly the case for the macroeconomic effects and the consequences on employment.

	One trillion yen	One billion dollars	Main cause	Source
Compensation for damages	8.3 (2018 ¥)	77 (2018 $)	Nuclear accident	(TEPCO 2018)
On-site costs for Fukushima Daiichi	8 (2016 ¥)	71 (2016 $)	Nuclear accident	(JCER 2017; TEPCO 2017)
Remediation of contamination and intermediate storage	6 (2016 ¥)	53 (2016 $)	Nuclear accident	(TEPCO 2017)
Economic losses from agriculture and fishing	2.4 (2012 ¥)	30 (2012 $)	Earthquake and tsunami	(MAFF 2012)
Effects on industrial production	N/C	N/C	Earthquake and tsunami	(METI 2011)
Tourism image costs (value is only for Fukushima Prefecture)	2.9 (2014 ¥)	27 (2014 $)	Nuclear accident	(Nohara and Narukawa 2015)
Increase in electricity production costs	> 11 (2014 ¥)	> 100 (2014 $)	Nuclear accident	(EIA 2015, 2016)
Macroeconomic effects	4% of GDP in Q1 2011		Earthquake, tsunami and nuclear accident	(Cabinet Office 2012)

Table 4.1. *Synthesis of the main costs from the Fukushima accident*

In addition, several financial and macroeconomic events had an important effect on the performance of the Japanese economy after March 2011, in particular the rapid fluctuations in the exchange rate of the Japanese yen with respect to all major currencies and the recovery plan for the Japanese economy that was decided on in 2012.

Concerning the economic consequences due to the Fukushima accident, significant costs were observed, arising from the loss of damaged reactors and their dismantling, the remediation of contamination in territories contaminated through compensation for the victims, but also due to the economic losses resulting from the decision of the Japanese authorities to temporarily cease the operation of all nuclear power plants in the country.

Table 4.1 provides a summary of the main costs of the Fukushima accident from the literature.

4.4.2. *Compensation for damages*

An initial estimate of the damages arising from the Fukushima accident was published on October 3, 2011 by TEPCO (2011). The consequences of the nuclear accident, based on the expenses paid as of that date, were estimated at 5825.5 billion yen (73 billion USD_{2011}). These expenses were broken down as follows. The losses of Fukushima Daiichi have been estimated at 964.3 billion yen (11.8 billion USD_{2011}). The damages were compensated in 2011 for the amount of 2,618.4 billion yen (32.8 billion USD_{2011}), and in addition, the estimated amounts of compensation for the following 4 years were estimated at 2,242.8 billion yen (28.0 billion USD_{2011}). This first estimate did not assess all the damages, particularly those related to the remediation of contaminated territories. This led to an underestimated result.

In mid-2018, the compensation amounts paid by TEPCO amounted to 8,310 billion yen (TEPCO 2018), an increase of about 70% compared to what had been initially planned. These costs could increase further over time, as compensation is paid.

4.4.3. *On-site cleaning and dismantling costs*

The post-accident management and work on the Fukushima Daiichi site, including decontamination and dismantling, will extend for at least 30–40 years after the accident (METI 2011). An initial on-site cost estimate was published by TEPCO on October 3, 2011 in TEPCO (2011). The additional cost of dismantling segments 1–4 was estimated at 964.3 billion yen (11.8 billion USD_{2011}).

According to the Board of Audit (Fukushima on the Globe 2015), TEPCO and the government had already spent 590 billion yen as of the end of 2014 (USD 4.93 $billion_{2014}$) in works on the Fukushima Daiichi site. In 2016, TEPCO updated the estimate of the cost of dismantling the destroyed Fukushima Daiichi reactors to 8,000 billion yen (TEPCO 2017).

4.4.4. *Costs to remediate the contamination*

MOE (2018) estimated the cost of the decontamination of contaminated territories at 2,900 billion yen (about 27 billion USD_{2018}) for operations that took place between 2011 and 2017. These costs are likely to increase as the decontamination operations progress. TEPCO (2017) gave a valuation of 6,000 billion yen (US$53 $billion_{2016}$) for the anticipated costs of management of contaminated territories and intermediate storage of the waste generated by decontamination operations (excluding the final waste storage).

4.4.5. *Economic losses from agriculture and fishing*

The economic effects on agriculture, fisheries and the agribusiness industry include direct damages to production and the decrease in agricultural yields; the increase in production, transport and transaction costs (placing goods on the market); the decrease in revenues and consequently farmers' incomes; the loss of capital value (land, equipment, intangible capital); the decrease in labor productivity; and the increase in health surveillance costs.

According to MAAF (2012), the total direct damages caused to agriculture, fisheries and forestry is estimated at around 2,400 billion yen (30 billion USD_{2011}). The fishing sector was the hardest hit by the disaster, with economic losses estimated at around 1,270 billion yen, the majority of which is attributable to damages to port facilities. The economic impacts on the agricultural sector are estimated at 910 billion yen, most of which concern the prefectures of Miyagi and Fukushima. The largest share of these economic losses is due to damage to arable land and agricultural facilities. There has also been a huge destruction of livestock, fruits and crops in the affected area. The majority of the economic losses in agriculture and fishing are therefore caused by the destruction of fields, boats, and infrastructure by the earthquake and tsunami. On the other hand, the fact that the economic losses persist is directly linked to the contamination of agricultural areas by the nuclear accident.

Following the nuclear accident, about 40 countries, including major importers, imposed restrictions on imports in the agribusiness sector from Japan after the accident. In 2012, most of them relaxed their import restrictions on the affected prefectures, but they maintained restrictions for the Fukushima Prefecture. Due to import restrictions, the value of Japanese exports of agricultural and livestock products has decreased: between April and December 2011, it fell by 40.9 billion yen (11%) (MAFF 2012). A decline in agricultural and fisheries exports was observed after 2010, while imports of agricultural, forestry and fisheries products increased (MAFF 2012).

It should be noted that some farmers and some agribusinesss companies in uncontaminated regions have benefited from positive effects due to increased prices and better production and sales opportunities. These benefits must be taken into account in the evaluation, and therefore deducted from the costs, in order to obtain an evaluation of the net costs of the benefits.

4.4.6. *Effects on industrial production*

The industrial sector, and in particular the automotive manufacturing sector, was severely affected by the accident, with a sharp decline in industrial production caused by interruptions in the supply chain. During the first month following the accident, industrial production in Japan as a whole fell by 15.5%, and by the end of 2014, industrial production had not yet returned to its values prior to the accident (METI 2011).

The impact on industrial production was greater in the prefectures directly affected by the tsunami (Iwate, Miyagi and Fukushima). In the Miyagi Prefecture, industrial production fell by more than 50% in the first 2 months after the disaster, and temporarily returned to pre-accident levels only a year after the accident. The trend in the Fukushima and Iwate prefectures is similar: a sharp decline (–50%) followed by a recovery. However, at the end of 2014, industrial production in the three prefectures remained about 8% lower than pre-incident levels.

4.4.7. *Effects on tourism*

Tourism in Japan has been severely affected by the earthquake, the tsunami and the Fukushima incident. The short-term effects on tourism have been significant, with a year-on-year reduction of 28% of tourists visiting Japan, from 8.6 million in 2010 to 6.2 million in 2011, and a comparable loss of income. According to a study carried out by (Cabinet Office 2012), the negative impact of the earthquake, the tsunami and the nuclear accident on the tourism industry is about 700 billion yen (about €5 billion).

At the national level, the Japanese tourism industry recovered quickly from the Tohoku disaster, and its impact has been much lower than initially expected. In 2014, about 11 million foreign travelers came to Japan, almost double the amount from 2011[6]. However, despite the recent spike in the number of foreign tourists in Japan, the tourism sector continues to represent a very small share of Japan's GDP, with international tourism revenues representing only 0.6% of GDP in 2016, a much lower level than in other developed countries such as France (2%) or the United States (1.3%).

The effects of the three simultaneous disasters on tourism were much greater in proportion in the Tohoku region (Cabinet Office 2012). Foreign visitors to the Fukushima Prefecture have fallen by 85% in 2011 and, 4 years after the accident, that figure is still 60% lower than it was in 2010. A similar trend can be observed in the Iwate and Miyagi prefectures. In 2017, the economic tourism sector of Fukushima Prefecture had finally returned to its pre-disaster level: the number of domestic Japanese tourists had reached 92% of the pre-disaster level in 2016, and the total number of international guests in Fukushima exceeded the pre-disaster level in 2017.

4.4.8. *Image effects on the agribusiness sector*

The decrease in imports of products from the agribusiness sector in Japan is partly related to the image effects to agribusiness. Thus, for example, the estimation of the damages caused to agricultural products by the Fukushima accident by Brachev and Ito (2018) is 2,568 billion yen (about €20 billion) in Fukushima Prefecture alone. It covers the damage of unsold products (planning and distribution restrictions) and the loss of value caused by image costs. This estimate suggests that, in comparison with the total value of 2,400 billion yen for the total direct damage caused to agriculture, fisheries and forestry in Japan (MAFF 2012), the image costs to the agribusiness sector after the Fukushima accident in Japan are potentially significant.

Aruga (2017) conducted a detailed study of the reaction of consumers to the agribusiness markets after the Fukushima accident. The study showed that the fear of the risk of radioactive contamination through the ingestion of food considered to be potentially contaminated had the effect of a real risk to the agribusiness market by reducing volumes and prices for consumption. According to Shigeru and

6 Many factors may have had an impact on tourist flows to and from Japan. The devaluation of the yen after 2013 has strongly encouraged the arrival of foreign tourists, mainly from Asian countries. The Japanese government has also made considerable efforts to promote Japan as a tourist destination and to reassure international travelers about the low level of radiation exposure outside a limited affected area.

Viet-Ngu (2019), farmers in the Fukushima prefecture lost from 13.1% to 18.9% of their sales due to the damage to the image of their products.

4.4.9. *Image effects on tourism*

A study has shown that the loss of the recreational benefits in the Fukushima Prefecture due to the economic damage caused by hearsay (image effects) amounted to about 2,850 billion yen (about €22 billion) between 2011 and 2014 (Nohara and Narukawa 2015). Decreases in tourist visits have been suffered, including in territories free from contamination. In this sense, in the city of Kyoto, the number of tourists in 2011 was 436,563, which is 55.6% less than the reference year 2010. According to Mingji et al. (2019), this reduction in the number of foreign tourists in Kyoto was influenced by harmful rumors (image effects) following the nuclear accident in Fukushima.

4.4.10. *Effects on the electricity generation system*

The Fukushima nuclear accident had a long-term impact on Japan's energy policy. Many thermal and nuclear power plants located near the epicenter, as well as transport and distribution infrastructure, were damaged after the earthquake and tsunami, and several nuclear reactors were removed from the grid for safety and maintenance tests. As a result, serious power outages occurred in 2011, with restrictions on electricity consumption and peak demand throughout the country, and the beginning of rotating power outages in the five most affected prefectures.

In October 2012, the new Nuclear Regulatory Authority (NRA) published the new procedures and regulations required to restart the nuclear power plants. In addition to a positive safety assessment by the NRA, it was necessary to obtain the agreement of the local government to restart the nuclear reactors. The first reactor to be put back into service after the Fukushima accident was in August 2015.

Between 2010 and 2012, the shutdown of nuclear reactors cost Japan 3,600 billion yen ($36.7 billion) (EIA 2015). At the end of 2014, Japan had spent an additional annual average of about $30 billion on fossil fuel imports in the 3 years following the Fukushima accident, and had experienced a record trade deficit of $116 billion (12.8 billion yen) in 2014 (EIA 2016). The total cost for Japan's electricity system therefore amounts to more than $100 billion.

In addition to the increase in electricity generation costs, nuclear power plant operators in Japan have been faced with additional costs for implementing the safety improvements necessary to restart nuclear power plants. WNA (2020) estimates these costs in the range of $700–$1,000 million per reactor, regardless of its size and age. The expenses incurred by electricity producers were estimated at $12.3 billion in March 2014; they were reassessed at $17.4 billion in March 2017.

4.4.11. *Macroeconomic effects*

The macroeconomic impact of the March 2011 disaster was significant. The country's real gross domestic product fell by almost 4% between January and March 2011 (compared to 2010), and Japan experienced a trade deficit due to increased imports (Cabinet Office 2012). According to the International Monetary Fund (IMF), GDP dropped by 0.7% in 2011, and the estimates for 2012 placed GDP growth at 2%, stimulated by reconstruction works following the destruction from the earthquake and tsunami.

4.5. Conclusion

This chapter has provided a brief overview of the various costs of a nuclear accident and the methods by which they can be assessed. An example addressing the costs of the Fukushima accident provided an illustrative example of this description. Finally, what should be the lessons readers learn from this?

First, the costs of nuclear accidents fall into several different categories: direct costs, indirect costs and macroeconomic effects. The first costs are directly linked to the damages suffered by the reactor and the radioactive contamination of the territory. The indirect costs, such as image costs and the effect of a nuclear accident on the mix of electrical power sources, depend on the decisions taken by consumers, distributors and the public decision-maker after a nuclear accident. The value of the parameters of the economic models corresponding to these indirect costs can be inferred from nuclear accidents that have already occurred in the world or from events of another nature, but recent methods seeking to identify individual preferences make it possible to improve these cost estimates.

Next, there is no single cost for a nuclear accident, but rather multiple costs that depend on the objectives pursued. The estimation of the cost of a nuclear accident thus only makes sense if the objective of the evaluation is specified, as summarized in SFEN (2015):

> The isolated use of a numerical figure, given out of context, can lead to a communication that can generate more controversy than progress

in terms of the knowledge of the subject. In the past, this situation has thus led to more rational debates being diverted or becoming more difficult. It can produce confusion between approaches carried out by different authors pursuing different objectives, significantly limiting their understanding and their usefulness. It is therefore important to clarify the analysis and results grids as much as possible.

4.6. References

APREF (2015). Risques nucléaires et transition énergétique. Paris, France.

Aruga, K. (2017). *Consumer Reaction, Food Production and the Fukushima Disaster: Assessing Reputation Damage due to Potential Radiation Contamination*. Springer International Publishing. New York, USA.

Bizet, R. and Levêque, F. (2017). The economic assessment of the cost of nuclear accidents. In *Resilience: A New Paradigm of Nuclear Safety*, Ahn, J., Guarnieri, F., Furuta, K. (eds). Springer, New York, USA.

Brachev, H. and Ito, F. (2018). Implications of Fukushima nuclear disaster for Japanese agri-food chains. *International Journal of Food and Agriculture Economics*, 2, 95–120.

Bromet, E.J., Havenaar, J.M., Guey, L.T. (2011). A 25 year retrospective review of the psychological consequences of the Chernobyl accident. *Clinical Oncology*, 23(4), 297–305.

Cabinet Office (2012). Knowledge note: CLUSTER 6: The economics of disaster risk, risk management, and risk financing – Earthquake risk insurance. World Bank Institute, Washington, D.C., USA.

CCI France Japon (2011). Dernières évolutions du régime d'indemnisation en cas d'accident nucléaire. Tokyo, Japan.

Cox, D.R. (1972). Regression models and life-tables. *Journal of the Royal Statistical Society*, 34(12), 187–22.

Drèze, J. (1962). L'utilité sociale d'une vie humaine. *Revue française de recherche opérationelle*, 23, 93–118.

Eeckhoudt, L., Schieber, C., Schneider, T. (2000). Risk aversion and the external cost of a nuclear accident. *Journal of Environmental Management*, 58(2), 109–117.

EIA (2015). Country Analysis Executive Summary: Japan. US Energy Information Administration, Washington, D.C., USA [Online]. Available at: https://www.eia.gov/international/content/analysis/countries_long/Japan/japan.pdf.

EIA (2016). Energy Policies of IEA Countries: Japan 2016 Review. IEA, Paris, France [Online]. Available at: https://www.iea.org/reports/energy-policies-of-iea-countries-japan-2016-review.

Euratom (2013). Council Directive 2013/59/Euratom of 5 December 2013 laying down basic safety standards for protection against the dangers arising from exposure to ionising radiation, and repealing Directives 89/618/Euratom, 90/641/Euratom, 96/29/Euratom, 97/43/Euratom and 2003/122/Euratom. Official Journal of the European Union, Aberdeen, UK.

Evans, J.S., Abrahamson, S., Bender, M.A., Boecker, B.B., Gilbert, E.S., Scott, B.R. (1985). Health effects models for nuclear power plant accident consequence analysis. NUREG/CR-4214, Rev. 2, Part I, ITRI-141, Albuquerque, NM, USA.

ExternE (2005). Externalities of energy, methodology. In *European Commission*, Bickel, P. and Friedrich, R. (eds). IER University of Stuttgart, Germany.

Fukushima on the Globe (2015). TEPCO and gov't spent nearly 600 billion yen to battle Fukushima disaster. FUKUDEN, Fukushima, Japan [Online]. Available at: https://fukushimaontheglobe.com/the_earthquake_and_the_nuclear_accident/4373.html.

Greenpeace (2013). Fukushima deux ans après : des victimes livrées à elles-mêmes [Online]. Available at: https://www.greenpeace.org/belgium/fr/rapports/1069/fukushima-deux-ans-apres-des-victimes-livrees-a-elles-meme/.

Hasegawa, R. (2015). Returning home after Fukushima: Displacement from a nuclear disaster and international guidelines for internally displaced persons. Preprint submitted on 5 March 2019. hal-02057182 [Online]. Available at: https://hal-sciencespo.archives-ouvertes.fr/hal-02057182/document.

ICRP (2007). The 2007 recommendations of the international commission on radiological protection. ICRP Publication 103. Ann. ICRP 37(2–4).

IRSN (2012). Guide d'aide à la décision pour la gestion du milieu agricole en cas d'accident nucléaire. Fontenay-aux-Roses, France [Online]. Available at: https://www.irsn.fr/savoir-comprendre/environnement/guide-daide-decision-pour-gestion-milieu-agricole-cas-daccident.

IRSN (2020). La recherche, étude probabiliste de sûreté. Fontenay-aux-Roses, France.

Ishii, K. and Morlans, S. (2013). La reprise des activités agricoles dans les régions contaminées après l'accident de Fukushima : un défi lancé aux agriculteurs. *Géographie et cultures*, 65–82 [Online]. Available at: https://doi.org/10.4000/gc.2891.

JCER (2017). Accident cleanup costs may rise to 50-70 trillion yen. Japan Center for Economic Research. Future of Energy and Environment Choices (Extra Edition). March 7, 2017.

Jones, T.D. (1981). Hematologic syndrome in man modelled from mammalian lethality. *Health Physics*, 41, 83–103.

Levy, H. (1994). Absolute and relative risk aversion: An experimental study. *Journal of Risk and Uncertainty*, 8, 289–307.

Louass, A. and Picard, P. (2017). Optimal insurance for catastrophic risk: Theory and application to nuclear corporate liability. Working Papers hal-01527478, HAL.

Louaas, A. and Picard, P. (2022). Optimal nuclear liability insurance 2019. *The Energy Journal*, 43(1) [Online]. Available at: https://hal.science/hal-01996648/document.

MAFF (2012). FY2011 annual report on food, agriculture and rural areas in Japan. MAFF.

Mariotto, A.B., Yabroff, K.R., Shao, Y., Feuer, E.J., Brown, M.L. (2011). Projections of the cost of cancer care in the United States: 2010–2020. *Journal of the National Cancer Institute*, 103(2), 117–128.

Mas-Colell, A., Whinston, M., Green, J. (1995). *Microeconomic Theory*. Oxford University Press, USA.

METI (2011). Chapter 4. In White Paper on International Economy and Trade 2011. Ministry of Economy, Trade and Industry, Tokyo, Japan [Online]. Available at: https://www.meti.go.jp/english/report/downloadfiles/2011WhitePaper/4-1.pdf.

Mingji, C., Qinglin, C., Yusuke, T., Hidehiko, K. (2019). A study on economic damage for tourism in Kyoto city by earthquake disaster. *Asian Journal on Hospitality and Tourism*, 13(2), 101–112.

MOE (2018). Environmental remediation in Japan. Ministry of the Environment, Tokyo, Japan [Online]. Available at: http://josen.env.go.jp/en/pdf/progressseet_progress_on_cleanup_efforts.pdf.

NCRP (2018). Implications of recent epidemiologic studies for the linear-nonthreshold model and radiation protection. National Council on Radiation Protection and Measurements, Commentary No. 27, Bethesda, USA.

Nohara, K. and Narukawa, M. (2015). Measuring lost recreational benefits in Fukushima due to harmful rumors using a Poisson-inverse Gaussian regression? *EconPapers*.

Nomura, T. (2014). Le droit japonais de la responsabilité des dommages nucléaires et son évolution après l'accident de Fukushima. *Revue juridique de l'environnement*, 39, 629–639.

OECD (2012a). Mortality risk valuation in environment, health and transport policies.

OECD (2012b). Japan's compensation system for nuclear damage. As related to the TEPCO, Fukushima Daiichi Nuclear Accident.

OECD (2015). Projected costs of generating electricity.

Oshima, K. (2012). Damage of Fukushima nuclear accident and new compensation scheme. Ritsumeikan University, Japan.

Rheuben, J. and Nottage, L. (2013). Now that the radioactive dust has settled: Resolution of claims from the Fukushima nuclear disaster. *Asian Dispute Review*, 2013(3), 126–131.

Rodríguez, G. (2007). Lecture notes on generalized linear models [Online]. Available at: https://grodri.github.io/glms/notes/.

SFEN (2015). Le coût des accidents nucléaires. Note de la Société Française d'Energie Nucléaire (SFEN), Paris, France [Online]. Available at: https://www.sfen.org/wp-content/uploads/2015/03/Le-coût-des-accidents-nucléaires.pdf.

Shigeru, M. and Viet-Ngu, H. (2019). Economic loss due to reputation damage: A new model and its application to Fukushima peaches. *Journal of Agricultural Economics*, 71(2), 581–600.

Szpiro, G. (1986). Measuring risk aversion: An alternative approach. *The Review of Economics and Statistics*, 68, 156–159.

TEPCO (2011). Management and finance investigation committee report.

TEPCO (2017). TEPCO integrated report.

TEPCO (2018). Records of applications and payouts for compensation of nuclear damage.

Viscusi (2005). The value of life. Discussion paper no. 517, Harvard John M. Olin Discussion Paper Series, Cambridge, MA, USA.

Wells, J. (1976). A guide to prognosis for survival in mammals following the acute effects of inhaled radioactive particles. *Journal of the Institution of Nuclear Engineers*, 17(5), 126–131.

WHO (2012). The great east Japan earthquake.

WNA (2020). Nuclear power in Japan.

5

Prospective Scenarios from the Present to 2050

Patrick CRIQUI[1], Rodica LOISEL[2] and Lionel LEMIALE[2]
[1]*CNRS, Université Grenoble Alpes, France*
[2]*IAE Nantes, France*

5.1. Global energy scenarios and the role of nuclear power

5.1.1. *A brief history of energy planning*

5.1.1.1. *1970s–1980s: before and after the oil shocks, the possibility of resource scarcity*

In terms of energy and the environment, 1972 represents a turning point from several points of view: it is the year of the publication of the MIT report at the Club of Rome entitled "The Limits to Growth"; the year of the first United Nations conference on the environment and development in Stockholm; and also the year in which the first strategic scenarios were prepared by Shell, created by its planning group seeking to analyze the fundamental trends of the energy markets. Later on in the 1970s, academic or applied research explored the conditions of development under the constraint of natural resources, particularly for energy.

The MIT report to the Club of Rome (Meadows et al. 1972) marks the dramatic beginning of "the finite world". This is true since the report's predictions about the scarcity of resources appeared to have been confirmed at the end of the following year by the occurrence of the first oil shock. Based on the "dynamics of systems" model by Dennis Meadows and his co-authors, the scenarios describe the possibility of a world that, if it continued down the same path, would lead to the exhaustion of

Nuclear Economy 2,
coordinated by Jacques PERCEBOIS and Nicolas THIOLLIÈRE.

the reserves of non-renewable natural resources, and consequently, a peak in industrial production and food rations by 2020, and then by 2050 a collapse of the world population[1] (see Figure 5.1).

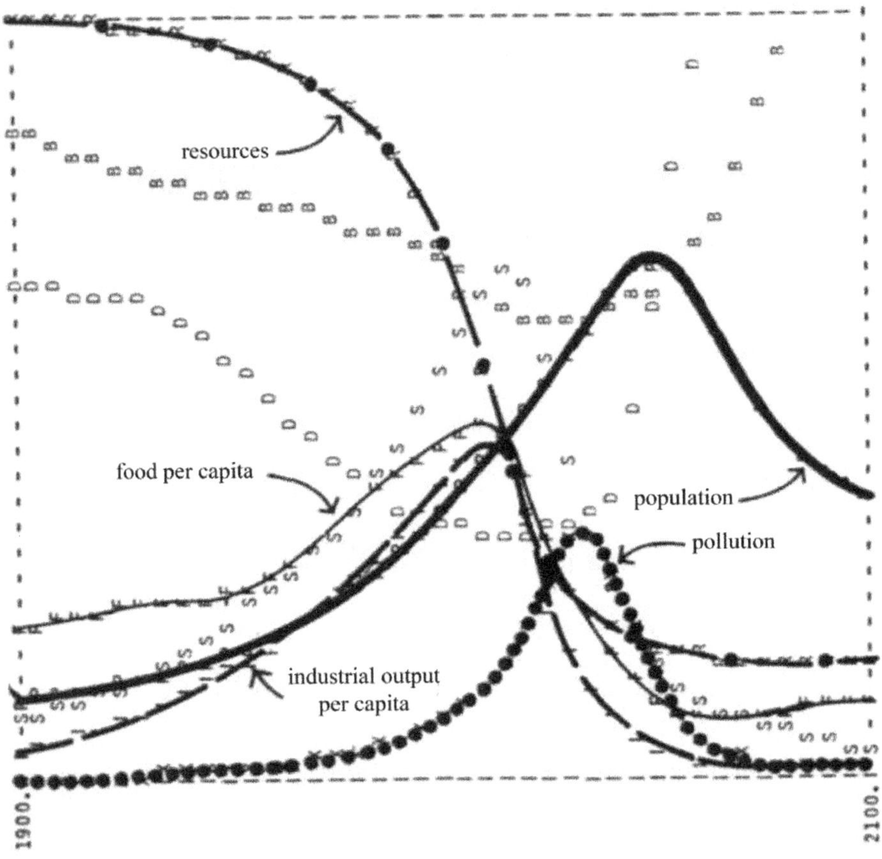

Figure 5.1. *The original diagram from the report "The limits to growth"*

During that same year (1972), the Planning Group of the Shell Oil Company, led by Pierre Wack, produced its first strategic scenarios. These predicted strong growth in oil consumption, the difficulty of supplies to keep up with demand, and therefore the possibility that a period of high oil prices would begin. For almost half a century now, the work produced by Shell would be a key point of reference in the petroleum

1 We will not enter into any discussion of this thesis here, which was extensively criticized or, according to some, later revalidated (Turner 2008).

world, both from the point of view of its methodology and the new visions that they would bring as the crises developed[2]. But the work of the Club of Rome will also elicit other responses of different types.

In the academic world, growth theorists would try for many years to respond to the Club of Rome by explaining the conditions for sustained growth to be possible. Of course, it seems sustainable today (Pezzey and Toman 2002). Inter-temporal optimal growth models allow us to explore the question of whether it is possible to replace the natural capital consumed through economic activity with artificial capital, built on the basis of an adapted investment. If this substitutability exists, then the limits to growth can be exceeded through what will be described as "weak sustainability"; otherwise, it will be necessary to either reduce the extraction of natural resources, a plan put forward as "strong sustainability", or to become exposed to the risk of collapse, as described by the Club of Rome.

Several research initiatives have taken place within a more applied field with the goal of exploring alternative energy scenarios that have the potential to overcome resource crises. These include the work of the WAES[3] in the 1970s, which would later produce the book *Energy for a Sustainable World* (Goldemberg et al. 1988), describing a world in which high efficiency levels in the use of energy makes it possible to satisfy the needs of developing countries, without running up against resource constraints. A similar perspective was developed at IIASA (Häfele et al. 1981), converging at the end of the 1980s with the views of the world energy conference (Frisch 1983; Grübler et al. 1996).

Based on a linear programming model for the intertemporal optimization of the energy system (Schrattenholzer 1981), this study explores three scenarios with contrasting profiles in detail: *High Growth*, *Middle Course*, and *Ecologically Driven* (see Figure 5.2). By 2100, global energy consumption would be less than half in the latter scenario than in the high-growth scenario: resource and environmental problems arise in radically different ways, including within the variants of each scenario which call for very different profiles for primary energies and particularly nuclear power: depending on the scenarios, the share of nuclear power in the global energy supply varies between 4% and 14%.

2 The experience from 40 years of using Shell scenarios is documented here: https://www.shell.com/energy-and-innovation/the-energy-future/scenarios/new-lenses-on-the-future/earlier-scenarios.html.

3 Workshop on Alternative Energy Strategies.

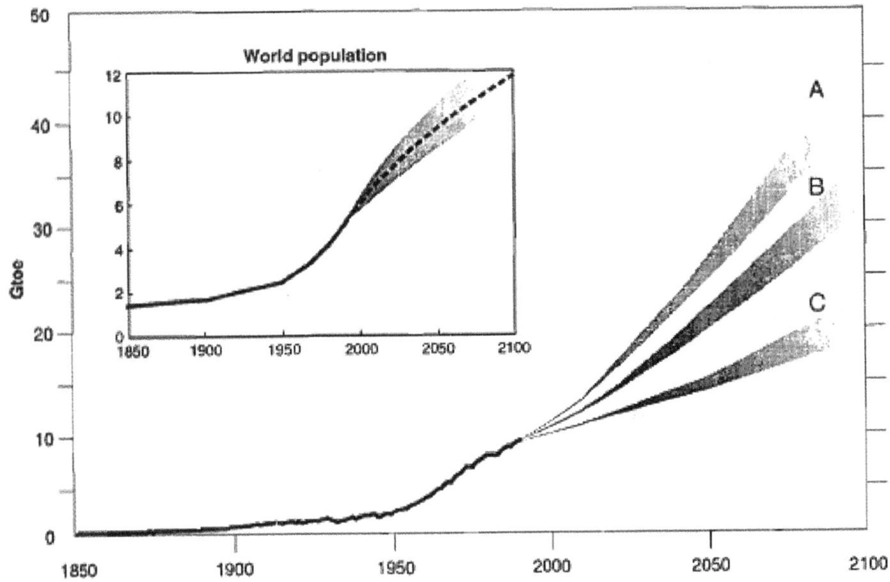

Figure 5.2. *The three scenarios from the IIASA-CME study*

5.1.1.2. *1990s–2000s: the climate issue and the motivating role of the IPCC*

Before 1988, concern over the climate emerged only gradually, though first appearing as early as 1975, when pioneers such as William Nordhaus – at the time in a residence at the IIASA – asked the question of emissions control (Nordhaus 1975). But with the creation of the IPCC, a new research paradigm took shape, in which the climate issue took precedence over that of resource constraints. It should also be noted that this happened 2 years after the oil counter-shock of 1986, which marked the beginning of a period characterized by a relative abundance of hydrocarbons.

The very structuring of the three IPCC working groups, created around the issues of climate science, adaptation and mitigation, created a favorable space for the deployment of efforts for developing projections and scenarios. This came despite the fact that even though the tools of the energy and economic sector had undergone a progressive maturation in previous years. Indeed, in the 1980s, two main lines of research allowed for both the improvement of energy models – supply optimization and demand simulation – and the development of economic models, which were either Keynesian, or general calculable equilibrium inspired by neoclassical principles.

Thus, at the beginning of the 1990s, when the issue of the modernization of climate policies entered the debate, four main families of models existed: optimization or simulation for energy models and general equilibrium models or Keynesian growth models for economic models. The toolbox used by the modelers for the climate economy would provide diversified resources that were possibly able to be combined. The combination of different modules gave rise to the so-called Energy–Economy–Environment (3E) models and the coupling with simplified climate models to the so-called integrated assessment models. The integrated assessment models consortium (IAMC) provided the documentation for the different models and coordinated their use in harmonized scenarios[4].

In particular, the results of the IAMs (in a broad sense) form the main inputs for the work by the IPCC Working group No. 3. And in this context, the framework for the construction of different scenarios was formed in 2000 with the creation of the special report on scenarios (Nakicenovic et al. 2000). The quadrant thus created allows us to clearly set the structural elements of the contrasting scenarios. The scenario tree that is defined (Figure 5.3) is rooted in a set of exogenous hypotheses regarding population, growth, technologies, resources, etc.

On the first axis, the vertical axis, the scenarios differ according to whether priority is given to economic growth objectives, or rather to environmental quality objectives. The second axis, the horizontal axis, compares differing levels of globalization, or the emphasis on regional and local elements. Each scenario therefore analyzes a "productivist" or "environmentalist" aspect (A/B) against an "openfield" or "hedged-in" one (1/2).

This scenario tree would inspire many works on energy projection; even today, 20 years later, it is still extremely relevant to the description of possible futures. In the following two IPCC reports of 2001[5] and 2007[6], the scenarios of the SRES would form the basis of the prospective works.

Models and scenarios were implemented during the 1990s and 2000s in many research projects, in particular in the framework research and development programs of the European Commission. These made it possible to conduct in-depth economic analyses of the costs of climate policies. This formed part of a diverse "epistemic community" with rich differences in the results it obtained in the models and its interpretative debates (Edenhofer et al. 2010). This community is organized in

4 Available at: https://www.iamcdocumentation.eu/index.php/IAMC_wiki.

5 Available at: https://www.ipcc.ch/report/ar3/syr/.

6 Available at: https://www.ipcc.ch/report/ar4/syr/.

international scientific forums such as the International Energy Workshop[7] and the Energy Modeling Forum[8] (Kriegler et al. 2014).

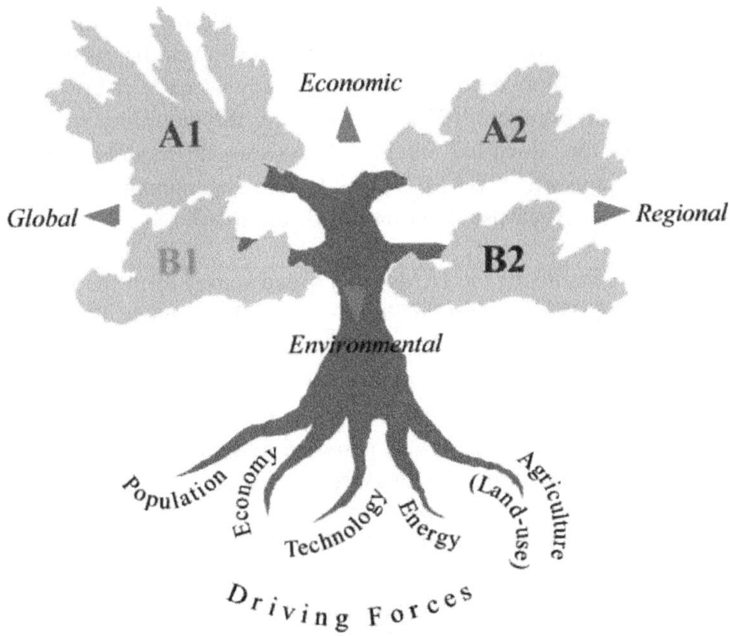

Figure 5.3. *The four scenarios of the IPCC report on emission scenarios (SRES). For a color version of this figure, see www.iste.co.uk/percebois/nuclear2.zip*

However, some economists would dispute these works over their lack of transparency and the difficulty in using their results to define effective policies to combat climate change (Pindyck 2013). While the question of validating the models and their results was raised, it appeared that the method of projecting based on past observations, commonly used for physical climate models, was difficult to implement in socioeconomic models.

The validation of the models and their results then requires a set of diagnostic indicators to be established to analyze the plausibility of the simulated changes (Kriegler et al. 2015).

7 Available at: https://www.internationalenergyworkshop.org/.

8 Available at: https://emf.stanford.edu/.

5.1.1.3. *Before and after the Paris Agreement, carbon neutrality and the establishment of a new paradigm*

Beginning in 2010, a new approach was defined within the IPCC. This approach sought to establish connections between the climate scenarios, built around representative concentration trajectories (RCP)[9] and the common socioeconomic trajectories (SSP)[10]. Four contrasting socioeconomic trajectories, which are then added together with a central scenario, are defined by the fact that they constitute more or less favorable contexts for the adaptation or mitigation of climate change (O'Neill et al. 2017). These form the basis of the modeling work of the Fifth IPCC Report (2014)[11], using the following nomenclatures: SSP1 = sustainability; SSP2 = central scenario; SSP3 = regional rivalries; SSP4 = inequalities; SSP5 = fossil development. In combination with the RCPs, these trajectories allow simulations of energy systems and in particular of the electrical system (Bauer et al. 2017) (see Figure 5.4).

However, in the context of the preparation of the Paris Agreement, other research efforts were carried out in parallel with the work of the IPCC Group 3. In some aspects, they constitute a paradigm shift for modeling in support of international climate negotiations. The Deep Decarbonization Pathways Project (DDPP; Sachs et al. 2014) was initiated by Jeffrey Sachs (then the economic adviser to Ban Ki Moon) and Laurence Tubiana (who would later be climate ambassador to François Hollande). The project mobilized roughly 20 national teams, including 15 countries from among the largest emitters[12] in order to build national strategic scenarios of deep de-carbonization (see Figure 5.5).

The methodological approach differs from that adopted with integrated models, since it aims not to obtain homogeneous results from a single model, but rather to identify specific trajectories, taking into account the resources, constraints and objectives specific to each country.

The approach had two interests: first, from the point of view of analysis, the diversity of situations is not erased by the use of a global model fed by common data; second, from a policy point of view, the perspectives thus constructed have the potential to be directly used by governments to build their deep decarbonization strategies. In this case, at the time when the study was launched in 2013, it was able to contribute to the preparation of national contributions before the COP21 in Paris and contributed in a modest extent to the success of the negotiations.

---

9 Representative Concentration Pathways.

10 Shared Socioeconomic Pathways.

11 Available at: https://www.ipcc.ch/report/ar5/syr/.

12 Italy was integrated into the project during its second phase.

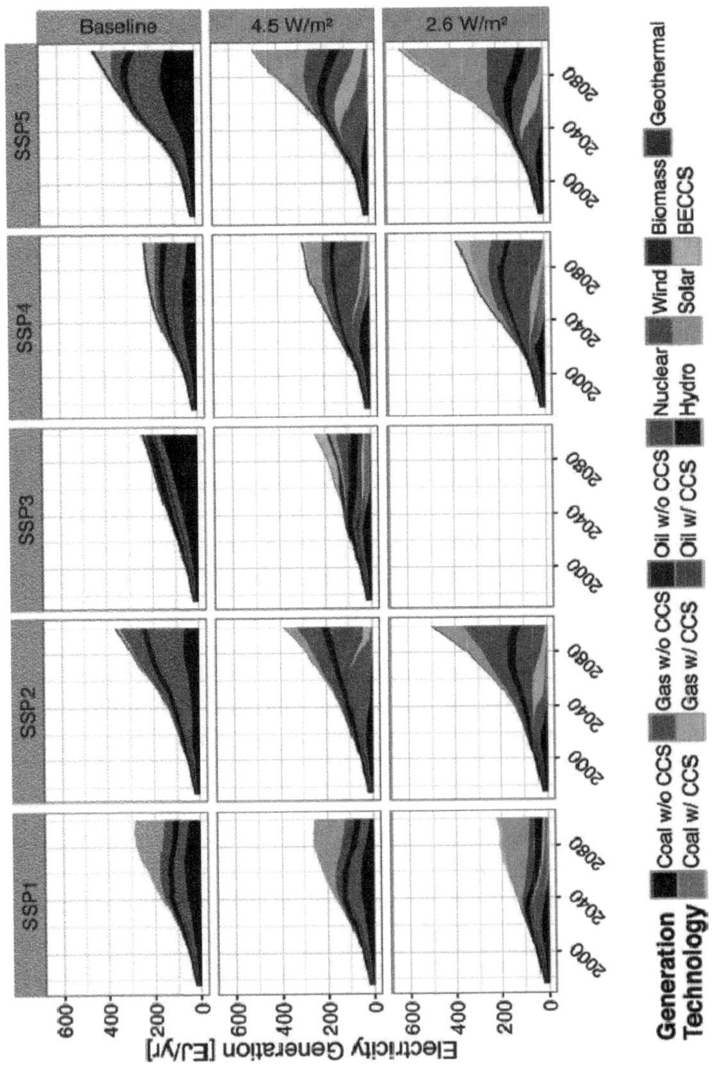

Figure 5.4. *The production of electricity in the combining of SSPs and RCPs (source: Bauer et al. 2017). For a color version of this figure, see www.iste.co.uk/percebois/nuclear2.zip*

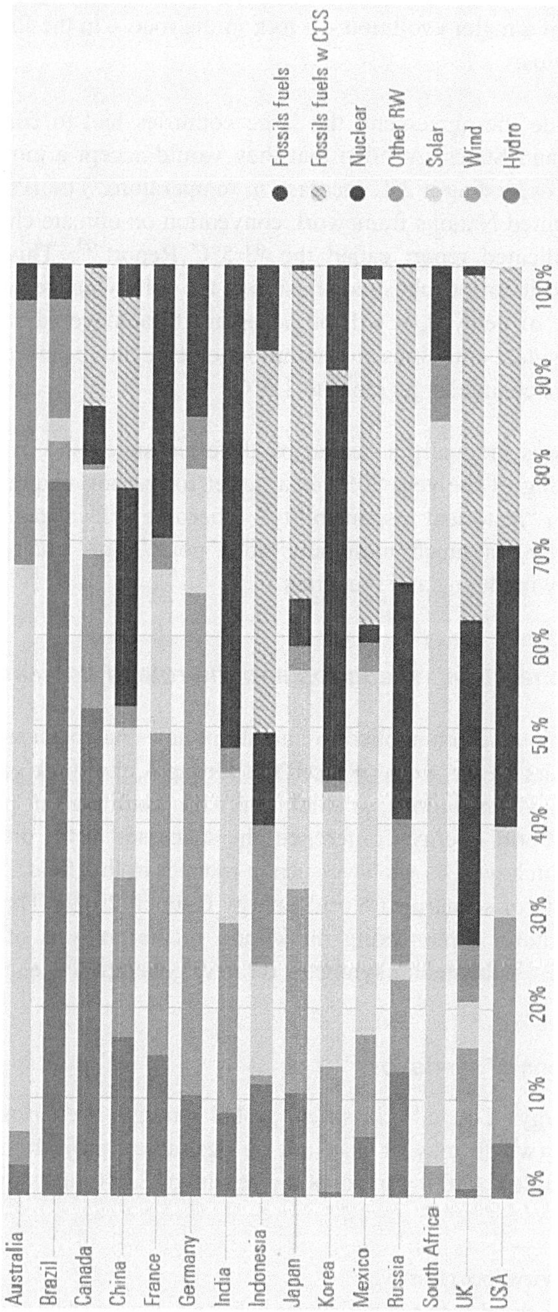

Figure 5.5. *Electricity production in the 15 countries of the Deep Decarbonization Pathways project. For a color version of this figure, see www.iste.co.uk/percebois/nuclear2.zip*

But even if its success must be put into perspective, the Paris Agreement above all planted the seeds of a major evolution – a fork in the road – in the formulation of energy-climate scenarios.

Indeed, to conclude the agreement, the large countries had to concede to the Alliance of Small Island States (AOSIS) that they would accept a more ambitious goal than that of not exceeding a 2°C increase in temperature. The IPCC was then called upon by the United Nations framework convention on climate change, tasked with preparing a dedicated report called the "1.5°C Report"[13]. This report was published in 2018 and its most robust conclusion is the following: regardless of the action scenario, early or delayed, it will be necessary to achieve carbon neutrality (i.e. the balance of residual emissions and anthropogenic negative emissions) around 2060 to meet a climate change target close to 1.5°C.

This conclusion was gradually accepted by different states and, particularly in Europe, the establishing of the year 2050 as a target for carbon neutrality – taking into account a certain "historical responsibility" – became a focal point of climate policies. By introducing a much more ambitious goal[14], this change upset the perspectives of energy modeling and scenarios.

5.1.2. *The "new generation" scenarios and the role of nuclear power*

While the academic research applied to climate policies has been structured and punctuated for 30 years by the work of the IPCC Group 3, the work carried out by companies, international institutions, or think tanks has continued in parallel up to today. In 2019, the World Energy Conference, the successor to the previous world energy conference which was, as we have seen, a pioneer in this field, conducted an inter-comparison study of scenarios (World Energy Council 2019). That same year, Resources for the Future, a pioneering think tank in the analysis of energy and environmental policies, published a synthesis on new generation scenarios (Newell et al. 2019).

5.1.2.1. *A comparison of scenarios*

The World Energy Council classifies global energy scenarios into three categories: plausible (which may or may not be desirable), projections (probable) and normative (desirable). And it compares the results of 2,040 of its own scenarios

13 Available at: https://www.ipcc.ch/sr15/.

14 In France, since 2005, the objective of the climate policy was the reduction of emissions by a factor of 4; carbon neutrality requires a reduction by a factor of 6.

with those of different authorities: oil companies (Shell, Exxon, BP, Statoil), international institutions (OECD-IEA, US-EIA, IRENA), think tanks (IEEJ, Enerdata) and finally, for the standard scenarios, the IPCC (see Figure 5.6).

A cursory analysis of the results identifies the following characteristics for each of the three families:

– In the "plausible" scenarios, global consumption in 2040 is between 16 and 20 Gtoe, starting from 14 Gtoe today; the dependence on fossil fuels remains strong, as they still represent between two-thirds and three quarters of the total supply.

– In the "projections", the gap is much smaller in terms of demand (between 17 and 19 Gtoe) and likewise, the results are more homogeneous with regard to the contribution of fossil fuels (between 70 and 80%); this is explained by the fact that plausible scenarios explore a wider field of possibilities.

– Finally, in the normative scenarios, the energy demand is again much more dispersed, given that starting from the current level of 14 Gtoe, it is established between a range of 10–17 Gtoe in 2040; likewise, the contribution of the different primary energies sees significant differences. However, in all cases, the share of non-fossil renewable and nuclear energies reaches or exceeds 50% of the total; this is explained by the fact that in all cases the first normative constraint is the limiting of emissions.

The inter-comparison made by *Resources for the Future* is based on another set of scenarios, though it includes common scenarios (Shell, IEA, EIA and IEEJ). One of the aspects highlighted in this inter-comparison is that of the sharp increase in consumption, and therefore in electricity production, found throughout all scenarios. Starting from the current level of 24,000 TWh, all scenarios exceed 34,000 TWh in 2040 and the highest scenarios reach almost 50,000 TWh by that date.

This is a constant in the new generation of global scenarios: electricity, which is often associated with hydrogen, constitutes a central vector for the decarbonization of energy systems. The growth of this energy vector is therefore stronger than that of the others.

According to the scenarios, by 2040 fossil fuels will represent from only 19% to 35% of the total electricity production, versus two-thirds in 2015, while nuclear energy, depending on the case, represents around 10% of the total, as shown in Figure 5.7.

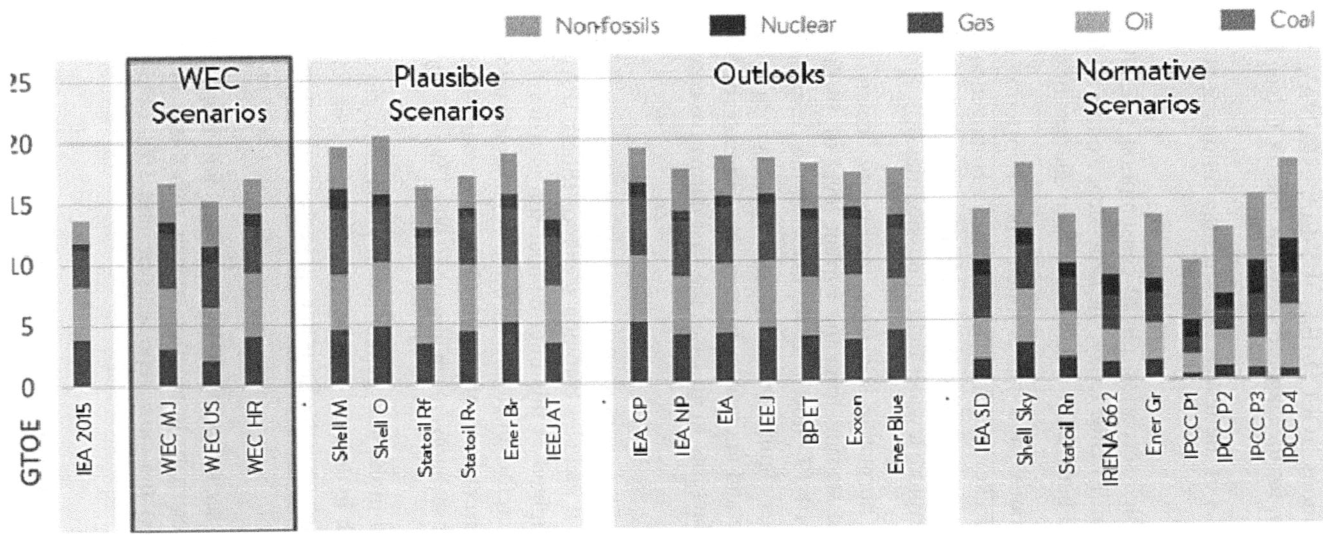

Figure 5.6. *Comparison of three families of scenarios by the World Energy Council (global energy scenarios comparison review report, published in 2019). For a color version of this figure, see www.iste.co.uk/percebois/nuclear2.zip*

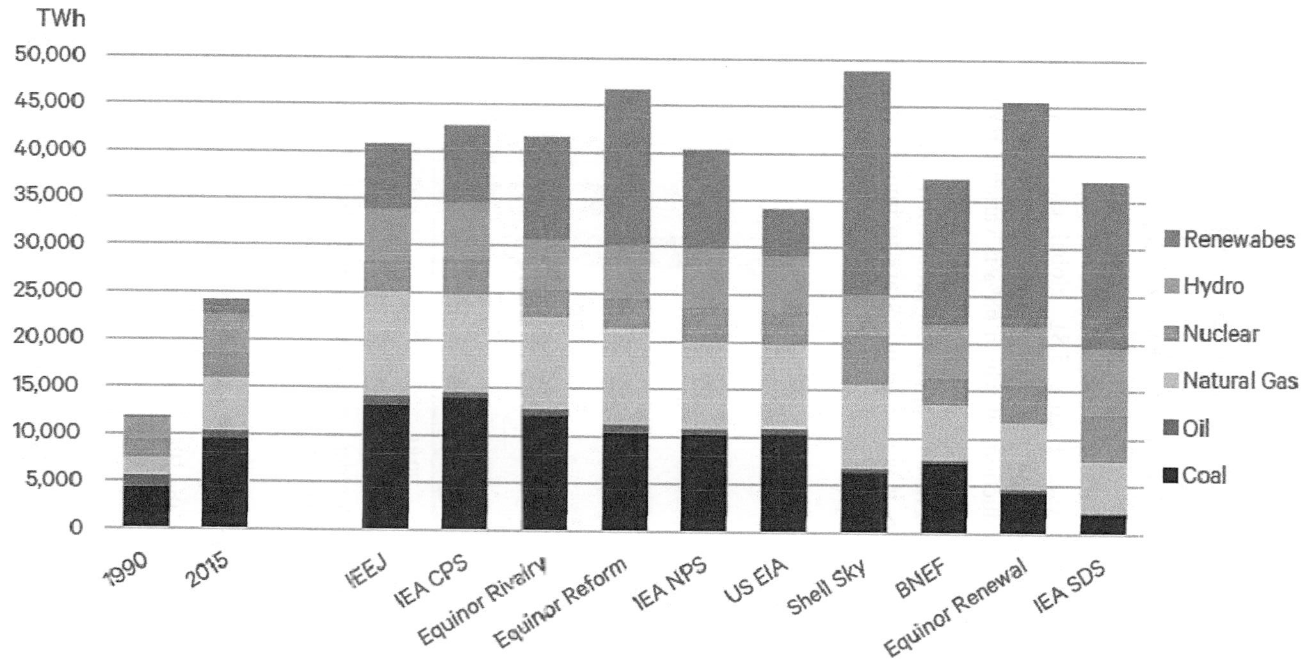

Figure 5.7. *Comparison of electricity production in the sample of scenarios from Resources for the Future. For a color version of this figure, see www.iste.co.uk/percebois/nuclear2.zip*

The overall diagnosis is confirmed by the most recent "new generation" global study by the IEA in the roadmap for net zero emissions in 2050 (NZE) (International Energy Agency 2021). This study is the first conducted by the IEA to explore the conditions for carbon neutrality in 2050. In the typology of the World Energy Council, it is clearly a "normative" scenario (1.5°C), built in contrast with two "projection" scenarios (Stated Policies Scenarios [STEPS] and Announced Pledges Case [PCA]).

In the NZE scenario (see Figure 5.8), global electricity production totals 56,000 TWh in 2040 and 71,000 in 2050. In this projection, renewable energies represent 88% of total production, of which 68% is variable renewables (wind and photovoltaic [PV] solar). Nuclear power would represent only 8% of the total amount, but this would still mean that current production would be doubled, which means that within 30 years it would be necessary to implement twice the nuclear capacity of today.

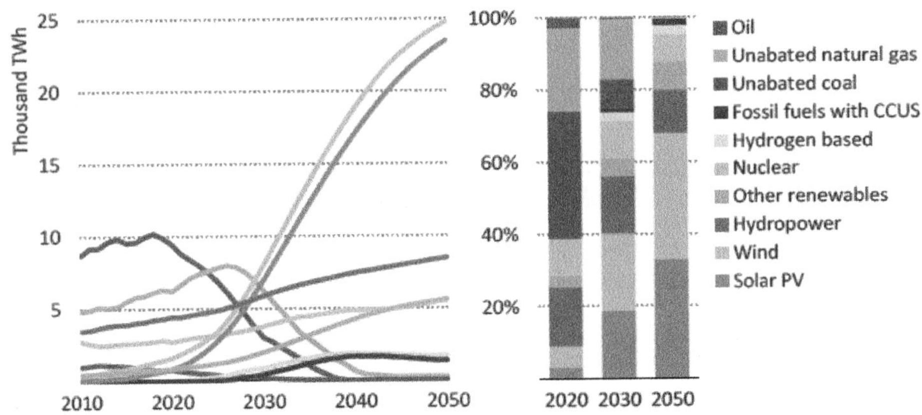

Figure 5.8. *Global electricity production in the IEA NZE scenario. For a color version of this figure, see www.iste.co.uk/percebois/nuclear2.zip*

5.1.2.2. *The challenges of modeling for carbon neutrality*

Whether in the work of the IPCC or in those conducted in France, such as by the Quinet Commission No. 2 (Quinet and Bueb 2019), the models for energy projection, with carbon neutrality, have been brought to their limits. In most cases, the objective can only be achieved under the following conditions: introducing exogenous and very restrictive demand assumptions (Grübler et al. 2018); assuming very significant negative emission rates (biomass with CO_2 capture and storage, direct atmospheric capture of CO_2); and ultimately assuming an extremely high price on carbon. Carbon neutrality therefore opens up new challenges for the creation of models and scenarios of energy and climate policies.

The difficulty for the exercises of modeled scripting is twofold:

– Carbon neutrality presupposes the transformations of energy systems and/or behaviors that are no longer incremental, but radical and systemic.

– To take into account the particularities of new renewable energies (physical potentials, variability), the models must represent spatial and temporary mesh below the usual mesh size, that is to say, the mesh at the national level in annual steps.

The modelers seized on these difficulties and integrated new technologies into their devices for describing the energy system (hydrogen, methane, storage, biomass with capture and geological storage of CO_2, etc.) and additional modules that would allow, for example, for the hourly variability of electricity supply and demand to be taken into account in systems with a high proportion of variable renewables. This is what allows them to build, despite all the uncertainties, coherent carbon neutrality scenarios.

However, the coherence of the models does not guarantee the feasibility and acceptability of the policies associated with them. And the choices for the trade-offs between the various drivers of decarbonization – actions on the supply or demand sides, inter-sector balances, the choice of the technological mixture – remain difficult to construct. Today, it is difficult to conclude that the models provide "proof" of the superiority of one strategic option over another regarding carbon neutrality.

5.1.2.3. *The uncertainties in the "technological trajectories"*

This relative ambiguity in the results of the long-term models can be explained by the uncertainties that have an obvious effect on the evolution of the various variables that influence the costs and the performances of the different technologies. For what has now been a long time, the scenarios modeled have taken into account technological dynamics, and in particular those associated with learning effects (Capros et al. 2012; Criqui et al. 2015).

The endogenization of technical progress in models and scenarios was essentially based on the concept of the "learning curve", using a factor of learning by use and, in some cases, learning by research (Kouvaritakis et al. 2000). In the past, these learning effects have played a very decisive role in the progress made by renewable energies (Wiesenthal et al. 2012). They are much more frequently discussed and even appear as negative for nuclear energy with exogenous factors of cost increase linked to the improvement of safety, but also endogenous factors that, particularly in France, are due to the desire to introduce an increasing share of design elements and components that are domestic in origin (Grübler 2010). In this field today, the reflection focuses on the analysis of the cost reductions to be expected from the effects of equipment size, which are opposed by the serial effects in the production of "granular" technologies based on small components (Wilson et al. 2020). The difficulties that have historically been faced in increasing the power capacity of nuclear power plants, and the potential gains from the mass production of equipment of a more modest size, offer a clear explanation for the current interest in small modular reactors (SMRs).

5.1.2.4. *Uncertainties about the robustness and acceptability of technological systems*

From the international prospective of energy today, it is possible to identify certain solid lessons and major areas of uncertainty.

Among the solid lessons is the fact that carbon neutrality will involve a very extensive electrification of energy systems and energy uses, likely greater than 50% of the end consumption. This will imply a strong increase in electricity production by volume, in the order of double current production by 2050. For this electric production, hydrogen will play a very important complementary role on all uses for which electricity is poorly suited (chemistry, industry and long distance transport) and taking into account the likely limitation of the potentials of bioenergies.

The areas of uncertainty, as we noted above, include those relating to the expected costs of the various decarbonized energy technologies. But it is still necessary to mention the uncertainties that arise first from the concern for the safety and robustness of electrical systems based on a very high proportion of variable renewable energies (68% in the NZE scenario of the IEA), and then from the limiting factor – one mentioned little if at all in the models – relating to the acceptability of the deployment on the proper scale of the diffuse energy solutions: wind and PV solar.

Given all these uncertainties, which are only considered in a very partial way in the models, the balance is now leaning significantly in favor of renewables as the main decarbonized source. But it remains to be seen whether the contribution of nuclear power in the future will remain limited to 10% of electricity production – which, as we noted earlier, already represents a doubling of production and capacities over thirty years, at a time when most of the existing power plants will have been decommissioned – or whether a more significant rebound can be expected.

5.1.2.5. *Small modular reactors*

If we examine the perspective of technological innovations, a new and potentially "disruptive" concept is currently emerging: the SMR. These reactors, with an output ranging from 10 to 300 MWe, are either derived through reducing the size of the architecture of the current second-generation reactors, or based on concepts taken from the fourth-generation reactors, such as molten salt reactors. Today, Russia has opted for the first option and China for the second, while both concepts are being developed in the United States. In France, EDF is preparing a project of the first type (NUWARD, 2 × 170 MWe), while research has been carried out for several years at the CNRS and IN2P3 on miniaturizable molten salt fast reactors.

This is indeed a potential rupture, because the SMR mark a complete change of perspective in the development of nuclear technologies. For the second-generation reactors, that of the power plants now in operation, the standardization and production series had allowed a certain mechanism for controlling costs. With EPRs, it has been apparent that the increase in size, which should have brought this cost control, has instead ultimately caused an increase in complexity, generating delays and additional costs, at least as of the present.

Conversely, the economic logic underlying the SMR projects is based on the concept of "granular technology" (in the sense given by the IIASA researchers), with small-sized equipment, the cost of which gradually decreases with the "learning effects" and large production runs. And in the last decades, this was exactly the case for wind turbines or photovoltaic panels, and it is also the case for batteries and electrolyzers today. This represents a shift from "economies of scale" to "serial effects".

This is clearly a bet on this technology. But faced with the climate imperative, it is obviously necessary to leave no solution unexplored. In addition to the large reactors, SMRs could offer new long-term prospects for nuclear power.

5.2. Modeling nuclear power in the energy transition

The goals of reducing GHG emissions by 2050 that many countries have set for themselves include the decarbonization of the electrical system through the massive deployment of renewable energies. For France, the energy transition law for green growth sets the share of renewables at 40% of electricity production by 2030 and 50% of the share of nuclear power by 2035 (Ministère de la Transition écologique et solidaire 2015). This significant increase in variable and intermittent energies leads to increased needs for flexibility in the electrical system, due to the inability of electricity to be stored on a large scale, and the need for supply and demand to be balanced instantaneously. For the past 10 years, this flexibility has been reflected in the variability of electricity prices on the spot market (Nicolosi 2010).

The flexibility of an electrical system can be defined as the ability to absorb an increase in uncertainty while ensuring the continuity and quality of the electricity supply (Goutte and Vassilopoulos 2019). This flexibility can be achieved by increasing the reliability of weather forecasts, promoting the geographical abundance of production sources, developing interconnections and energy storage or smoothing demand through the development of intelligent devices (Johnstone and Haščič 2012). However, these solutions pose real technical and economic challenges, as shown by the RTE report on the mass-scale development of renewables (RTE 2021). Finally, there is the option to ask for greater maneuverability by conventional technologies, particularly nuclear power plants.

In this section, we focus on this last alternative, particularly on how electrical scenarios anticipate the need for flexibility and include participation by nuclear reactors, taking into account their technical constraints. We combine two complementary modeling perspectives: one long-term perspective that determines the trajectories of the energy mixture under the constraints of political objectives, and another short-term perspective to simulate the electricity market under the technological constraint of flexibility.

Complementarity emerges from the time scale of each vision, in particular by combining the annual step for planning installed capacities up to the year 2100 (the POLES model) with the half-hourly step for the dispatching of power plants over the very short-term (EcoNUK model). Several obstacles can be identified upon considering the operations of the French electricity mixture, one dominated by nuclear power and renewables, such as the functioning of nuclear power in load monitoring mode, the need for fast reactors by 2050 and the possible trade-offs between power and energy to achieve the objective of reducing nuclear power in the electricity mix.

In section 5.2.1, we describe load monitoring for nuclear power plants and then in section 5.2.2 present the methodology used; we discuss the results of the reference scenario in section 5.2.3 and conduct a sensitivity analysis in section 5.2.4; finally, in section 5.2.5, we present the consequences on the organization of the sector before offering some final conclusions (section 5.2.6).

5.2.1. *Load monitoring in nuclear power plants*

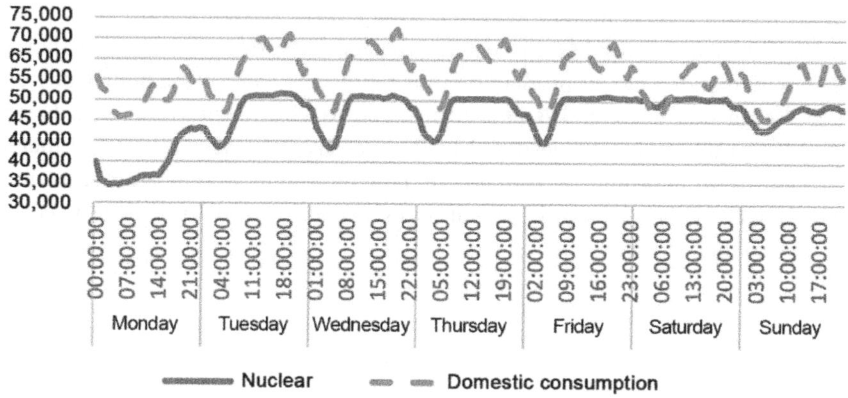

Figure 5.9. *Profiles of nuclear production and domestic demand during the first week of 2018 (source: the authors, based on data from RTE)*[15] *. For a color version of this figure, see www.iste.co.uk/percebois/nuclear2.zip*

With high fixed costs and low variable costs, nuclear power generation traditionally operates as a base supply. But in electrical systems with very high shares of nuclear power, such as in France (>70% of the mix), the nuclear plant network has been contributing to the balancing of the system since the mid-1980s, using three types of production modulation: primary and secondary calibrations, and load monitoring (Muniglia 2017). The first two consist of small power variations (±7% of the nominal capacity), while load monitoring represents the variation in the power of the reactors to adjust the supply to the demand for electricity in real time (IAEA 2018). This type of flexibility is provided by the nuclear operator (EDF) at the request of the system operator (RTE), and the adjustment is carried out manually by decreasing the power of the reactor to a previously agreed level[16] (Morilhat et al. 2019).

French nuclear production appears to be correlated with domestic demand (Cany et al. 2018a), since this production is adapted to its daily (day/night), weekly (working days/weekends) and seasonal variations (the peak summer production of the nuclear power plant, which in 2018 was 40 GW, represents 66% of the peak winter production, at 60 GW).

---

15 Available at: www.rte-france.com/eco2mix consulted on 30/01/2021.

16 For a detailed description of the process of piloting the production of reactors, see OECD/NEA (2012), Muniglia (2017) and IAEA (2018).

Figure 5.9 illustrates the ability of the nuclear plant network to absorb large variations in demand. The weakness of the alternative means of production used during demand peaks, combined with the strong variations during off-peak periods such as holidays and at night, requires the nuclear network to ensure the balance in supply and demand. Nuclear production thus represents 84% of the hourly power on average during this week and is able to respond to significant variations: decrease (–13.5 GW on the night of Tuesday 2 to Wednesday 3 January 2018) and increase (+12.7 GW in the morning of January 3). This represents a 22.4% decrease in the nominal power of the reactors in operation on January 2 at 17 p.m., followed by a 19% increase the next day.

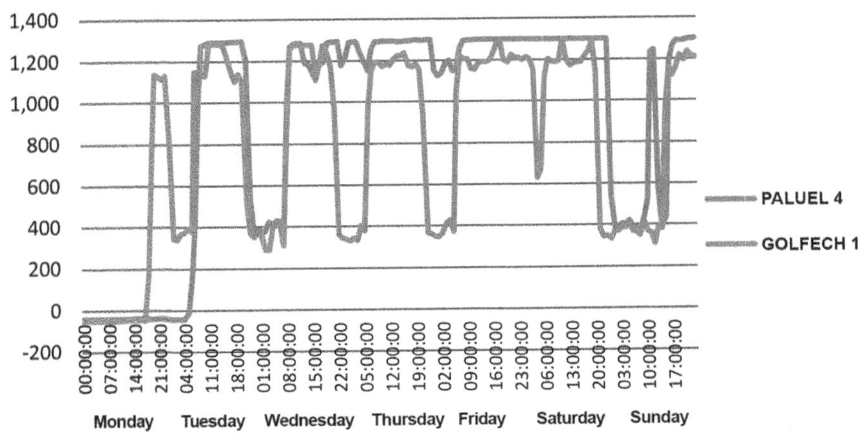

Figure 5.10. *Hourly production of Goldfech 1 and Paluel 4 during the first week of 2018 (source: the authors, based on RTE data[17]). For a color version of this figure, see www.iste.co.uk/percebois/nuclear2.zip*

Not all reactors participate in load monitoring in the same way. Out of the 56 reactors in operation at 5:00 p.m., only 21 adjust their production to demand between 5:00 p.m. and 4:00 a.m. Of these, two reduce their production by a factor of between 15 and 30% of the nominal production (Np), 15 by a factor of 50–80% Np, and finally, four reactors reduce their production by more than 80% Np; all then return to their initial production level. These adaptive behaviors are observed throughout the week (see Figure 5.10 for an illustration of the behavior of two reactors: Golfech 1 and Paluel 4).

17 Available at: www.rte-france.com/eco2mix consulted on 30/01/2021.

A load monitoring cycle consists of adjusting the power such that output is reduced, from 100% to a maximum safety level of 20–30% of the nominal power, followed by the return to the initial level (OECD/NEA 2011); in the graph shown in Figure 5.10, the parameters of the Golfech 1 and Paluel 4 power plants are around 30% Np. Variations can be frequent and low in amplitude, known as light cycles, or less frequent but more extensive, known as deep cycles (AREVA-EDF 2009; EDF 2013; IAEA 2018). During the first week of January 2018, Golfech 1 used five deep cycles while Paluel 4 used three, though with several low amplitude power variations (~15%).

According to Muniglia (2017), 48 French reactors (out of 60 initially) have been designed to participate in load monitoring, but only 40% of the network is currently used to ensure these operations (Cany et al. 2018a). These power variations (ramping) are in the order of 1% Np/Min, but the allowable speed is roughly 5%/min (OECD/NEA 2011). For safety reasons, the maximum budget allowed by a pressurized water reactor (PWR) reactor license is 12,000 deep cycles over 40 years, with a maximum frequency of one to two events per day (EUR 2012; Persson et al. 2012).

Below we identify three types of power adjustment cycles: shallow, medium and deep, which represent an amplitude of 20%, 40% and 70% of the nominal power of a reactor, respectively. The budget of cycles allowed over a year, by assumption, is 200 for deep cycles (100–30–100), 300 for medium cycles (100–60–100), and 3,000 for light cycles (100–80–100) (Ludwig et al. 2010) for comparable orders of magnitude. Exceeding these amounts will lead to consequences in terms of reactor wear and ultimately on their service life.

Since flexibility is anticipated in the design of the PWR reactor itself, its operation does not entail additional costs within the limit of the maximum budget allowed by the law. Thus, any event planned in advance allows the organization of the uranium supply cycle and for the reactor to switch to flexible mode without additional cost[18] (EDF 2013). However, certain costs may be incurred during the maintenance of components at the end of their technical lifespan, such as control rods and water and steam management systems (IAEA 2018), which are nonetheless in line with the aging process of a reactor.

Numerous studies have analyzed the consequences of the operation of the nuclear network in load monitoring mode in a context of high penetration of intermittent energies. These relate to the additional costs, the possible benefits and the number of cycles, as well as stoppages and starts of the reactors (Bruynooghe et al. 2010; Cany et al. 2017, 2018b; Ponciroli et al. 2017; ANL (Argonne National

18 Unlike planned cycles, unanticipated power variations can lead to additional fuel costs (in the order of 24%, as given by Persson et al. (2012)).

Laboratory) 2018; Bertsch et al. 2018; IAEA 2018; Jenkins et al. 2018; Mantripragada and Rubin 2018; OECD/NEA 2011, 2012, 2015, 2019; WNA 2020). Some studies look at the critical thresholds of nuclear and renewables in the mix of power sources, from which the variations in nuclear power that are induced are no longer compatible with safety constraints (Cany et al. 2017) or cause the nuclear load factor to decrease sharply (up to 20% in NEA (2012)). Despite these losses in market share, the flexibility of the nuclear network can reduce the cost of integrating renewables) (Ponciroli et al. 2017; Jenkins et al. 2018), and in some cases can increase the profitability of the nuclear network by reducing its exposure to price reductions, or by replacing power plants with little flexibility.

Our study complements these approaches by focusing on the profile of load monitoring to characterize the use of reactors that operate in flexible mode, following the trajectories of power source mixture imposed by political and climate objectives. In the event of excessive use of nuclear load monitoring, the study establishes a normative vision of how reactors must adapt to intermittent energies, and areas for reflection on the management of the nuclear plant network.

5.2.2. *Methodology*

The study of energy scenarios is based on two complementary categories of modeling: investment models (energy planning models), which determine the trajectories of the energy mix over the long-term, and the programming models for the plants (plant dispatching), which give a detailed representation of the technologies over a fine-grained temporal scale. This last category makes it possible to take into account the variability of intermittent energies, and to evaluate the *system* costs generated by intermittency (Göransson and Johnsson 2009; Van den Bergh and Delarue 2015). Certain studies explicitly address the need for flexibility by coupling between the two types of models (Times-Antares [Alimou et al. 2020]; POLES-Eucad [Despres et al. 2017]; POLES-Eutgrid [Allard et al. 2020]).

The electricity supply mix in France has been the subject of numerous optimization or exploration exercises, such as ADEME (2016, 2017), Negawatt Association (2017, 2022), as well as RTE (2017, 2021). Despite sharing a common base, these studies have major differences in terms of their assumptions and constraints, which makes it difficult to compare them in order to build a common vision on the transition in France (Shift Project 2018). In this section, we develop a modeling method to understand the consequences for nuclear policy planning scenarios of an integration of the operating modes of the nuclear network depending on whether it is considered as a base, semi-base or very flexible technology. These assumptions influence the management of technologies, both conventional and renewable, and impact the use of gas turbines and greenhouse gas emissions, the wear and tear of reactors, and future investment choices.

The methodology consists of coupling an investment model and a dispatching model. The first of these models describes the trajectory of the electrical system in a way that is compatible with political objectives; the second tests its technical and economic feasibility over the very short-term (30 min).

POLES (Prospective Outlook on Long-term Energy Systems) is a model for simulating the global energy system by 2050, using exogenous hypotheses concerning demographic and economic developments, for each of the 57 major countries or regions of the world (Criqui and Mima 2012, 2015). It is a bottom-up model that represents the oil, gas, coal, biomass and hydrogen markets by endogenizing the variables that characterize energy consumption, transformation, production and prices. Electricity generation is based on 41 types of technologies, 10 of which are renewable at input. The model thus chooses a very significant spatial and technological disaggregation, hindering its ability to provide high temporal resolution, since the simulation is carried out over a time period limited to 2 h per series of 12 time segments and once per summer and winter. The nuclear power plant network simulated in POLES is part of an investment strategy determined under the twin constraints of climate concerns and cost-effectiveness.

EcoNUK (Economic dispatching of NUClear reactors) is a model for the programming of power plants which, over a given year, optimizes the electrical system as informed by the POLES model through the installed capacities and the marginal costs of operation. EcoNUK describes 12 types of power plants and optimizes the functioning of the deregulated electricity market, minimizing the annual cost of the electricity system under the constraint of demand (see a detailed description in Loisel et al. (2018)). The model describes the interaction between technologies on a fine scale of 30 min, while reproducing the final state described with POLES and adding a number of constraints related to the maneuverability of the reactors such as their availability to provide flexibility. Despite the ability of all the installed reactors to operate in flexible mode, only two-thirds of the network operates in load monitoring mode, and the rest in base mode, in order to integrate the constraint of stabilizing the production of the reactors at the end of the uranium supply cycle (Morilhat et al. 2019).

Other technological constraints are included, such as the minimum operating threshold, the maximum annual load factor and the gradients of variation in the power of the reactors (0.1% Pn per half hour when the reactor is operating in base mode, and 5% Pn per half hour in load monitoring mode).

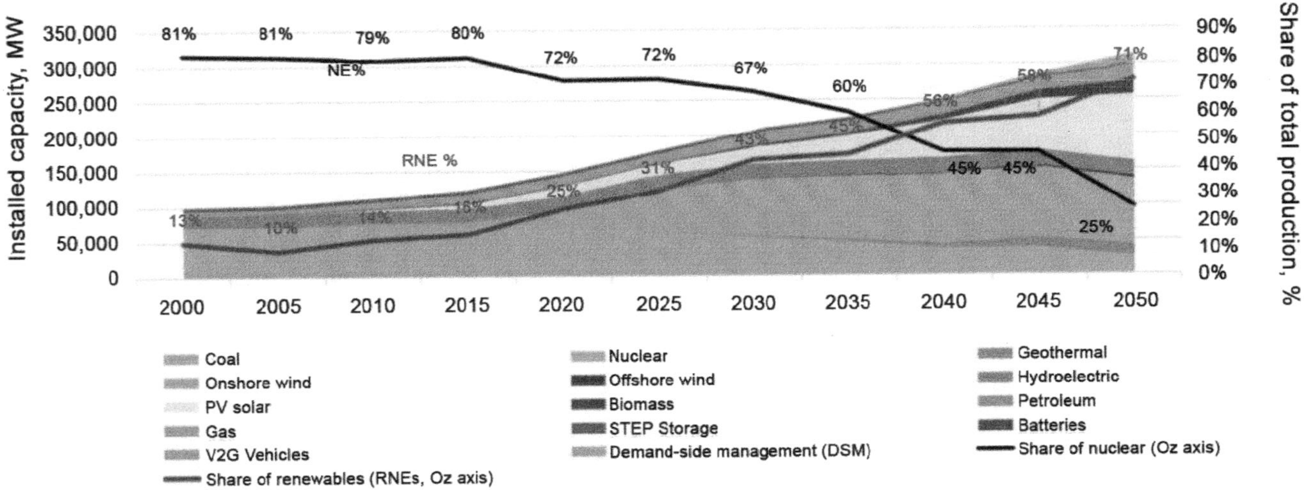

Figure 5.11. *Scenarios obtained using the POLES model for France under the constraint of a maximum temperature rise of 2°C at the global level by the year 2100. For a color version of this figure, see www.iste.co.uk/percebois/nuclear2.zip*

The scenario simulated by POLES is a global energy scenario for 2100 with the constraint of complying with the objective of limiting the temperature increase to 2°C. For France, intermediate steps are included such as the shares of nuclear and renewables in 2035. Figure 5.11 represents the electricity mix in France until 2050 and indicates a final mix formed with 25% nuclear and 71% renewable, of which 66% is at fatal input.

The steps of the coupling are as follows:

– The political objectives inform the shares of nuclear and renewables in the electricity mix to be achieved in 2050, as well as the CO_2 emissions.

– The POLES model simulates the year 2050 and transfers the resulting mix of electricity generation and storage capacities, as well as supply, demand and export-import flows.

– The EcoNUK model optimizes the given electrical system, constrained by the annual capacity factors from POLES, and returns sub-hourly factors that make it possible to study the nature of the flexibility of nuclear reactors, according to the cycles of variation of the power.

5.2.3. *Technical and economic analysis of nuclear reactors in scenarios up to 2050*

The results of the coupling confirm the conclusions of other studies, showing that the use of a model with high temporal granularity such as EcoNUK tends to amplify the need for flexibility compared to a model that aggregates variable energy inputs and technological constraints (Poncelet et al. 2014; Alimou et al. 2020). This need for additional flexibility made clear both by a modification of the controls of the technologies to the supply/demand balance, and as well as by an increased use of nuclear load monitoring.

We note a greater use of gas and the use of stored energy with the half-hourly step model, which seems to indicate tensions at certain periods, and also a punctual curtailment of renewable energies and hydroelectricity compared to POLES. With constant linear energy between the two models, the difference in the results (the highlighted column of Table 5.1) indicates a significant need for gas-based energy, at 14.3 TWh, and a total elimination of renewable energies of 7.3 TWh over the year, of which 70% is solar energy.

The additional needs in terms of gas and use of stored energy are generated in particular by the constraints of the ramping of nuclear reactors when providing positive flexibility (increases). Table 5.2 shows that when the speed of ramping increases, gas production decreases. In the event of negative flexibility (downward supply adjustment), all adjustable power plants participate, and when all possibilities are saturated, the system removes excess renewable energy. Alternative tests to the mix from the POLES model show that, for example, a mix of 57% nuclear and 43% renewables could avoid any removals and ensure the free supply–demand balance without any tension on the market in terms of installed capacities and the provision of flexibility.

Indeed, beyond the share of variable renewables, it is the unwavering supply of electricity that must be taken into account when planning the mix and the proportion of schedulable power plants. The intermittent capacity credit, calculated as the firm capacity of renewables to provide energy at any time of the year (Tapetado and Usaola 2019) totals 15 GW in 2050 in our model, which strongly influences the design and operation of the nuclear network, namely the supply of energy (GWh) and capacity (GW). Figure 5.12 shows the projections for 2050 from POLES, and the results obtained with EcoNUK.

Successive tests around the reference scenario show that the type of nuclear adjustment cycles evolves with the increase in renewable capacities: the more the share of intermittent sources increases, the deeper and longer the load monitoring, with longer plateaus. Conversely, the frequency of light cycles, specific to residual demand with low variability, is reduced. In 2050, the number of deep cycles decreases by 14% of the cycle budget allowed by the license (see section 5.2.1), and subsequently leads us to test the robustness of the mix when faced with the changes in the share of power sources (nuclear vs. renewables). We can thus characterize the frequency of the adjustment as a function of the ramping, the minimum safety threshold of the reactors and the sharing of the plant network between operations in base mode and in load monitoring mode.

The nuclear strategy described by POLES results in an annual utilization rate of 75% for the entire network. In EcoNUK, this rate applies to the nuclear network as a whole and the model breaks it down endogenously by type of use. Thus, the capacity factor is 78% for the flexible group and 71% for the group operating at a base level.

The results show that the system requires more flexible reactors than those at the base level, because they must compensate for both the intermittency of renewables and the rigidity of the base group. The group of plants operating at a base level is replaced by renewables during most times due to a massive flow of energy and a significant capacity credit (15 GW), which is higher than the base group (8 GW).

Scenario 2050 by model	POLES			EcoNUK	
	Capacity	Production	Capacity Factor	Production	Production
Technology	MW	GWh	%	GWh	Δ
Nuclear	24,030	159,697	76%	159,697	0
Coal	1,369	6,613	55%	-	6,613
Oil	4,330	-	0%	1	-1
Gas	16,301	17,578	12%	31,912	-14,333
Hydroelectric	21,851	60,927	32%	56,988	3,939
Onshore wind	91,929	192,293	24%	192,218	74
Offshore wind	4,355	14,558	38%	14,490	68
PV solar	94,240	106,376	13%	103,119	3,258
Other renewables	37,289	81,484	25%	81,485	0
Total	**295,694**	641,175	25%	639,910	1,265
Removal of demand	3,979	1,304	4%	1,394	-90
Storage (STEP + CAES + Batteries)	7,599	1,647	2%	10,475	-8,828
Import Interconnection, MW	27,000			27,000	0
Export Interconnections, MW	34,000			34,000	0
National Demand, GWh	471,478			471,478	0
Net exports, GWh	54,611			54,611	0
Losses, GWh	115,085	17.9%		115,085	0
Nuclear/Production	25%			25%	0%
Nuclear/Demand	34%			34%	0%
Renewables/Production	71%			70%	1%
Intermittent renewable/Demand	66%			66%	0%

Table 5.1. *Results of the POLES and EcoNUK models; variation Δ = POLES-EcoNUK*

5.2.4. *Sensitivity analysis*

In the reference scenario, the number of nuclear cycles is greater for deep load monitoring than for moderate load monitoring, as illustrated in Table 5.2 (baseline column).

These results are conditioned by dividing the nuclear network between flexibility and base operations (with 2/3 operating flexible and 1/3 base), a minimum safety threshold of 30% of the nominal power, as well as an adjustment speed of 5% per half hour throughout the year.

By testing the influence of each of these constraints on the behavior of alternative technologies (gas, storage, demand-side shaving), we obtain a measure of the contribution of the nuclear network to the flexibility of the system.

	Baseline	**Ramping**			**Minimum power threshold**		**Share of the network in the Base**	
	Ramp 5% Pmin 30% Base 33%	**1%/** 30 min	**10%/** 30min	**20%/** 30min	20%	50%	10%	50%
Cycles								
Light	**161**	267	195	246	152	179	183	140
Medium	**95**	88	67	74	90	92	92	81
Heavy	**228**	11	296	311	229	231	225	239
Capacity factors								
Base network	**71%**	66%	68%	65%	73%	64%	62%	73%
Flexible network	**78%**	81%	80%	81%	78%	82%	77%	78%
Dispatching								
Gas, GWh	**31,912**	36,834	29,817	28,683	31,818	33,049	29,751	33,609
Storage, GWh	**10,475**	12,099	9,175	8,039	10,494	10,231	9,690	10,993
Erasure of renewables, GWh	**4,695**	7,606	3,747	3,392	4,598	5,775	3,417	5,745
Max Implicit cost, €/MWh	**639**	639	639	621	639	639	639	639

Table 5.2. *Sensitivity of the results to the ramping rate, the safety threshold and the proportion of the networking operating at base levels*[19]

The adjustment speed: when the adjustment speed of the reactors increases (from 5% per half hour in baseline reference at 10% or 20%), the reactors substitute for the supply from gas and energy storage mechanisms. This also reduces the removal of renewable energies. These improvements in integrating intermittent sources together with a decrease in storage suggests that the latter provides a flexible means of support, first for technologies with adjustment constraints, such as nuclear

19 The type of cycle indicates the amplitude of the load monitoring: light cycles have an amplitude between 0% and 20% of the nominal power (100%–80%–100%), moderate cycles between 20% and 40% (100%–60%–100%) and deep cycles between 40% and 70% (100%–30%–100%).

reactors, and only provides support for renewable energies in its second stage. Increasing the power gradient rates ultimately allows for the supply-demand balancing of electricity to be done more easily, as shown by the maximum implicit cost on the market (known as the shadow price), which represents the cost of producing an additional unit of electricity and reveals the tension on the market. Conversely, a reduction in the speed of adjustment of flexible reactors delays activity on alternative technologies. Nevertheless, more reactivity should also involve additional (fixed) costs for the nuclear operator in case that a streamlining or upgrading is performed, since the number of deep cycles increases and exceeds the budget authorized by the license by more than half the total amount.

The minimum safety threshold: the results show a low sensitivity to the minimum power threshold. Indeed, any modification of this parameter leads to low variations in flexibility capacity, more specifically, +1,600 MW of flexible capacity (20% minimum power threshold) and –3,200 MW of flexible capacity (50% minimum power threshold). However, the effects match the expectations. Indeed, an increase of this threshold to 50% decreases the flexibility of the reactors and delays the demand for flexibility on other technologies, leading to an increase in gas-based production as well as an increase in the elimination of energy generated from renewables.

The distribution of the network between flexibility and base: the relationships that emerge from the tests show that the higher the proportion of the network used in base mode, the greater the increase on pressure on flexible reactors, with an intensifying supply of deep cycles. On the other hand, a decrease in the base share increases the degree of flexibility of the system, thus reducing the adaptation needs for other technologies. The number of deep cycles is also reduced, since the load monitoring needs are distributed over a larger number of reactors.

To summarize, it has been demonstrated that the system is particularly sensitive to changes in the power modulation speed of the reactors. This is reflected in the adjustment market through the supply of gas and unused power, and on the nuclear operator through the change in the budgeting of deep cycles. At the extreme end of the possible tests, the value of nuclear flexibility is called into question when (1) all reactors are used in non-flexible mode; and (2) a reactor is made unavailable for technical or managerial reasons for an entire year.

If the nuclear network operates entirely in base mode (non-flexible mode), the system encounters more difficulties in integrating renewable energies and leads part of these flows being eliminated (9.7 TWh). Figures 5.12 and 5.13 illustrate the voltage levels at specific points resulting from the lack of nuclear flexibility, events that focus on time slots where there is a high penetration of solar energy: an inflexible network requires more gas, storage and eliminated energy from renewables. Conversely, a flexible nuclear network facilitates the integration of solar and wind power at the rate of its adjustment speed (Figure 5.13). We can thus note a kind of removal of nuclear power through the offering of flexibility. However, we mention less the notion of removal for conventional energies, but more often the decrease in the capacity factor or the loss of market shares. If a flexible reactor is out of service throughout the year 2050, the system seems to be able to meet the demand. The available and operational capacity in 2050 demonstrates a sufficient level without additional voltage on the market, as indicated by the stable level of the implicit cost of the capacity value to satisfy an additional MWh. The effects on other technologies are the increase in the volume of electricity from gas (700 GWh) and more removal of renewables (+400 GWh). The hourly profile of gas-nuclear substitutions shows a preponderance during periods without sun, with a strong seasonal increase in winter. About 4,000 h are needed to mobilize more easily scheduled mechanisms to compensate for the prolonged shutdown of the reactor, totaling 46% of the hours over the year. However, the volume of gas-powered thermal production is less than the volume of nuclear electricity taken out of the exchange; in other words, the substitution is not effective at the ratio of 1 MW of gas to 1 MW of nuclear. Other nuclear reactors in base and flexible modes are called on to provide power, which increases their capacity factor in this capacity stress test (from 71% to 80% for the base group, and from 78% to 79% for the flexible group).

In terms of power, the maximum gas requirement over the year totals 1,740 MW, which demonstrates that removing a 1,000 MW reactor requires more schedulable capacity, due to the constraints of ramping the residual network which sometimes encounters more difficulties to ensure the same flexibility. This also results from the modeling assumptions of the network that distributes fatigue and stress uniformly within the reactors, which implies that one less reactor reinforces the cycles performed by the other reactors, in frequency or amplitude.

The sensitivity of the electrical system to energy flows tests the ability of the system to meet demand while respecting the gas budget described by POLES. Recall that the distribution of energy sources in France in 2050 according to POLES is made up of renewables (at 71%), nuclear (at 25%) and gas (at 3%). In EcoNUK, this scenario (Baseline) leads to a greater use of gas and use of stored energy (see section 5.2.3). Assuming a constant gas flow between the two models (a gas test), we then obtain feasible solutions for a similar but not identical mix: renewables (70%), nuclear (28%), gas (3%) and storage (2%). Thus, conserving gas flows from POLES requires more flexible nuclear energy, that is, 18 TWh more compared to POLES.

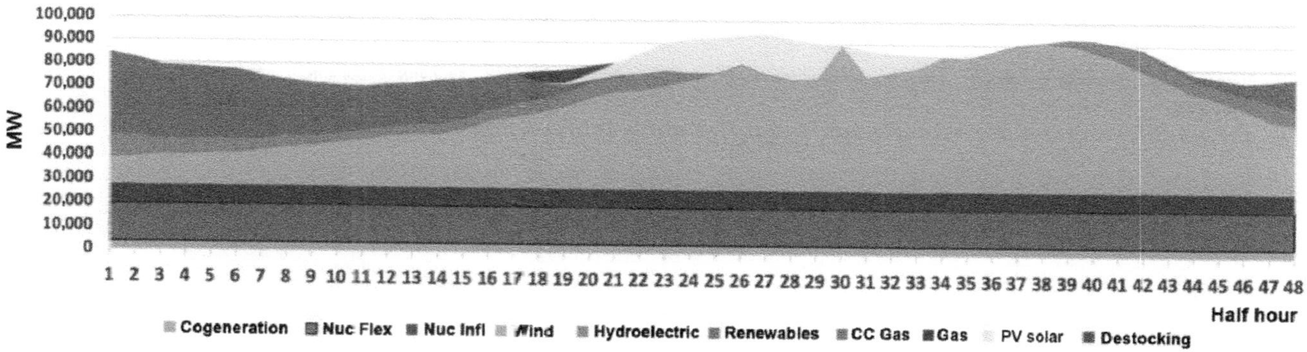

Figure 5.12. *Comparison of the operation of a base group and a flexible group at a 2:3 ratio. For a color version of this figure, see www.iste.co.uk/percebois/nuclear2.zip*

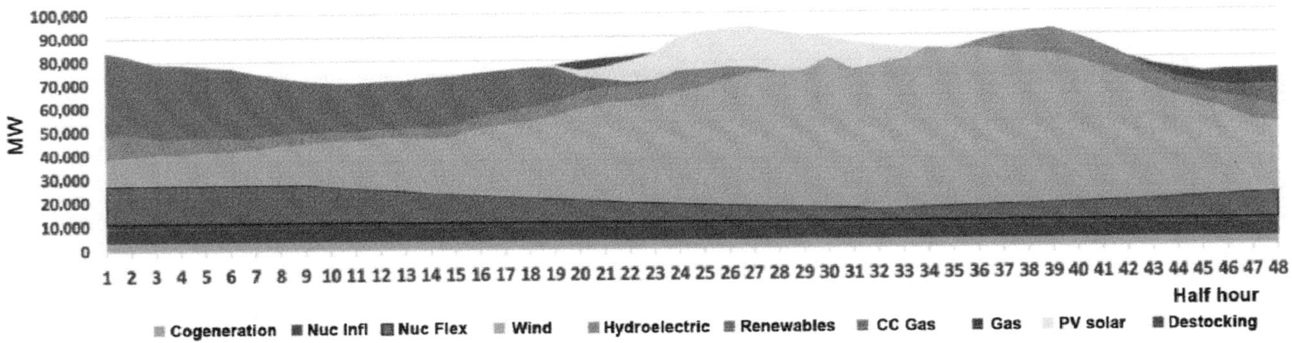

Figure 5.13. *Comparison of the operation in which the flexible group makes up 2/3 and the base group 1/3 for a day in January 2050. For a color version of this figure, see www.iste.co.uk/percebois/nuclear2.zip*

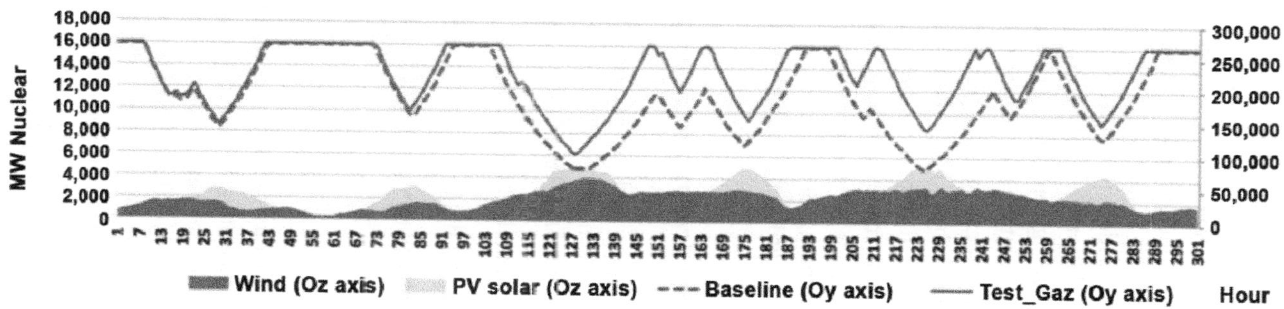

Figure 5.14. *Flow of nuclear energy from the flexible group, results from EcoNUK in 2050 in the scenario of Constant_gas (Test_Gas) and constant nuclear scenario between the two models (Baseline). Nuclear production in load monitoring mode over 6 days in August 2050. For a color version of this figure, see www.iste.co.uk/percebois/nuclear2.zip*

Figure 5.16 shows that on the selected days in August, the system needs more negative flexibility during sunny hours, and positive flexibility outside these time periods. When the gas is kept constant between the two models, more flexible nuclear power is delivered to the market (Test_Gaz, red curve) compared to the Baseline scenario, scenario in which it is the nuclear power that remains constant between the models (the dotted nuclear supply, Baseline).

The political choice to be made will ultimately be between more nuclear energy or more gas, assuming that the criteria for choosing between these energy mixes, such as energy independence, greenhouse gas emissions, and security can be comparable. The gas can be synthetic in nature (with a low emission coefficient when it is renewable, such as biogas or from green methanation) or it can be natural gas generating methane and carbon, but equipped with a carbon capture-storage system, even if at present the forecasts for the development of this technology remain low. When the system makes more use of nuclear power, the need for flexibility places higher demands on the network, that is, more shallow and deep cycles, which leads to a rethinking of the design of nuclear technology in terms of reactivity, in particular with regard to the budget of authorized adjustment cycles.

5.2.5. *Organizational, technical and technological perspectives*

In addition to the specificities and the precise accounting of the nuclear assets, our tests reveal structural tendencies for nuclear planning in the context of the energy transition. We can summarize them in three essential points:

– the management of uranium by anticipating the capacity factor;

– nuclear planning following the two structuring vectors: power and energy;

– the provision of centrally planned nuclear flexibility in the management of the network.

5.2.5.1. *Anticipation of the flow of uranium*

A determination of the capacity factor that is as accurate as possible in year N-1 for the following year is essential to the organization of the uranium supply campaign for the reactor, since the rate of use of the fuel is directly linked to this campaign. This will subsequently affect the inventory of nuclear waste flows and inventory, and more broadly, it forms part of the national waste management strategy. Therefore, using less gas and more nuclear power, as in the Test_gaz scenario, leads to high nuclear capacity factors (90% for the Base group and 82% for the flexible group), levels much higher than the historical rates (~75%). Among the objectives of the National Plan for the management of radioactive materials and waste (PNGMDR 2018) is the stabilization or even the reduction of plutonium inventories. France's strategy, based on the monorecycling of plutonium, implies an increase in its inventory at a rate of about 8 tons per year (for the nuclear network of 2018) (Thiollière et al. 2018). According to Cour des comptes (2018), the control of the plutonium inventory involves multi-recycling this element. The balance between inventories and the recycling rate is not guaranteed if a significant number of reactors do not use MOx. This highlights the interest in anticipating the evolution of the inventory for the organization of the retirement and storage infrastructure, such as storage pools that could become saturated in the near future, from 2025 to 2030.

5.2.5.2. *Nuclear planning, power-energy arbitrage*

The multiannual energy programming provides for the closure of 14 reactors by 2035, at the rate of one reactor per power plant (PPE 2020). The 2023 PPE could potentially refute this hypothesis. The closure of one reactor per plant, instead of entire plants, seeks to limit the economic and social cost in terms of jobs, regional dependence on industrial sites and a balanced electricity supply by region. In view of the political and social arguments, the debate on the future landscape of the electricity mixture could lean in favor of maintaining power plants with a decrease in their outputs with respect to their historical volumes, which would make it possible to achieve the objective of diversifying the mixture, but by reducing their capacity factors.

The significant need to provide flexibility in the EcoNUK model translates into a pressing need for capacities for several hundred hours a year, especially in winter and at night, with wide seasonal variations for short periods, such as during the day. Consequently, reducing the share of nuclear power could also be done by modifying the scheduling of the number of reactors to be shut down, since any closure will lead to additional stress on the remaining reactors. This calls for additional solutions in anticipation of periods with a lack of nuclear capacity, such as storage periods, demand management and green gas, which are given as almost perfect substitutes for nuclear power; these will make it possible to avoid the infeasibility solutions obtained with our market model, facing the two strong constraints of decreases in nuclear capacity and natural gas.

5.2.5.3. *Provision of centrally planned flexibility*

The functioning of the electricity market is governed by a decentralized design of "stacked" power supplies from plants on the basis of their marginal cost, though EDF's nuclear electricity offer is formed within a complex group in terms of technological diversity and the dispatching of power plants. More particularly linked to flexibility, reactors require new routines given intermittent renewals, new safety standards and human resources trained in oversight and monitoring (IAEA 2018). Like any new process for work and organization, the time required for training can be long, so as to break the inertia inherent in traditional work routines and arrive at new procedures to allow the reactors to have greater maneuverability (Tillement and Hayes 2019).

As nuclear power is strategic by nature and demanding in terms of safety standards, the management of the plant network is subject to close coordination at three levels: between the nuclear operator EDF, the safety authority ASN and the system operator RTE. These three levels influence the decision to operate in load monitoring mode, at very different times from when it is supplied:

– The Nuclear Safety Authority (ASN) has the primary role in the nuclear operation in view of its responsibilities for the authorization and regular compliance monitoring of power plants with respect to safety regulations; this implicitly covers the provision of flexibility in compliance with the design of the power plant according to the license, as part of a broader strategy for radiological monitoring.

– The nuclear operator, EDF, is directly responsible for the operation of the reactors, in the monitoring to account for the cycles carried out and in making decisions regarding the load monitoring process (IAEA 2018).

– The decision to monitor the load is taken at the request of RTE, which establishes the technical conditions for adjusting the frequency and the speed of reactivity following the national or even EU network codes. The cooperation between the two players is essential in ensuring that the needs of the network are in compliance with the monitoring capacity of the power plants, which leads over time to protocols that facilitate the provision of flexibility.

In conclusion with regard to the organization of nuclear flexibility, the process of establishing load monitoring arrangements is far from being a flexible and fast protocol as determined through real-time adjustments of supply to demand; on the contrary, it requires the prior organization of the steps for coordination between the regulator, the nuclear operator, the equilibrium manager and the operator in charge of waste management, with the support of scientists (IRSN, CEA, etc.), civil society and public decision-makers. Nuclear flexibility has become an almost completely rigid process due to its deadlines for adaptation, regulation and innovation, and these aspects often fall outside the construction of long-term scenarios that must include the feedback of stakeholders more broadly in order to validate the ability of the electrical system to be flexible.

5.2.6. *Recommendations for the section*

The mass-scale implementation of renewable energies within the electricity mixture, which is necessary to respect the objectives of the energy and climate transitions, will lead to increased needs for flexibility. In electrical mixtures composed mainly of renewables and nuclear energy, the variability of residual demand should be reflected in this technology.

Through the use of a very short-term dispatching model, this chapter explores the way in which nuclear power can meet the system's need for flexibility. The results of the simulations show that load monitoring facilitates the integration of renewables by reducing the amounts of energy lost from these sources, and in this way contributes to reducing the *system* cost. Nevertheless, a high proportion of intermittent production requires the production network to become increasingly adaptable, thus exceeding the cycle budget defined by the operating license, and resulting in an acceleration of the decommissioning of power plants or an increase in maintenance investments.

Faced with difficulties in terms of costs, safety and planning that can result from a distribution composed primarily of intermittent and nuclear production capacities, we explore alternatives to nuclear flexibility in order to facilitate the integration of renewables: the supply of the current nuclear network with a flexibility-energy co-product in the form of H_2 or steam; a new design for the reactors in terms of adjustment speed; and the deployment of SMRs.

1) Electricity-steam-hydrogen cogeneration: the advantage of nuclear power is its ability to provide both electricity and steam, as the steam could then be used for industrial or residential purposes through a municipal heat distribution network. The feasibility of the heating option has already been evaluated for several power plants in France and its technical and economic viability is highly dependent on density savings to absorb the nuclear supply and reduce the cost of the heating network, and on the proximity between the nuclear power plant and the urban center that would make use of it (Thiollière et al. 2018).

When providing flexibility, the reactor operates at a reduced load rate and undergoes the pressure described in the form of adjustment cycles; instead, a function at the maximum load rate could make it possible to produce electricity continuously and to use the energy produced that exceeds the needs of the network for the production of hydrogen. The production potential of H_2 appears significant, estimated at 1 Mt H_2 per year in Scamman and Newborough (2016) and up to 1.2 Mt per year by Cany et al. (2017), while still maintaining a large nuclear fleet in 2050 (between 40 and 60 GW).

2) New reactor designs: the various scenarios that test the ways in which to integrate renewables while maintaining a significant share of nuclear power rely on a rapid response by the reactors in terms of flexibility. Two adjustments will be necessary with respect to historical load monitoring, the speed of power adjustment and the reactor license in terms of the number of deep cycles that are allowed. These adjustments would make it possible, in the event of excessive use of nuclear flexibility, to avoid refitting and early maintenance, or even the closure of the plant due to exceeding the cycles allowed by the license, with additional costs or sunk costs in the event of a closure. The number of additional cycles estimated in our scenarios increases from the current level of 200 per year to about 250 in 2050.

Another technological improvement, and one which has not been tested in this part, concerns the waiting time between two cycles to stabilize the xenon oscillations, a time currently regulated at about four hours.

Finally, the reduction of the minimum safety threshold from 30% to 20% does not seem to have a significant impact on the system, due to a low nuclear capacity in 2050 compared to the renewable energy production network.

3) SMRs, which are described in section 5.1.2.5, represent an additional source of flexibility in terms of providing easier frequency control and locations on isolated sites, while also responding to the model for the transition to decentralized governance. The advantages of these include their shorter response time, less restrictive minimum thresholds and a shorter flexibility forecast plan, which would make SMRs a solution well suited to provide support for the intermittent nature of renewable sources (ANL 2005). France is diversifying its technological choices by supporting research on SMRs through the NUWARD project, a joint effort between the CEA, EDF, Naval Group and TechnicAtome groups, with funding in the amount of €470 million, as an investment for the ecological transition.

5.3. The study "Energy Futures 2050" by RTE (2021)[20]

RTE (*Réseau de transport d'électricité*, or "electricity transmission network") is a company responsible for the operation, maintenance and development of the France's high and very high voltage network. RTE guarantees the balance between insertions into and withdrawals from the electrical network at all times and can alert the public authorities in the event of a risk of medium and long-term blackout.

20 This section was written by Jacques Percebois and Nicolas Thiollière.

In 2022, RTE is owned by EDF (50.1%), Caisse des Dépôts et Consignations (29.9%) and CNP Assurances (20%).

In response to a referral from the government, in 2019 RTE began a study entitled "Energy Futures 2050", publishing it in Autumn 2021, whose goal is to provide a detailed analysis of the evolution of the electrical system[21].

The report produced by RTE it is divided into 12 chapters, specifying the objective and the scope of the study. The assumptions regarding energy consumption and production are specified in order to be able to produce scenarios for the evolution of the energy and the mix of electricity sources. Specific focal points are presented, which concern Europe, the security of the energy supply and hydrogen. Finally, the details of the economic and environmental analyses are put forward.

5.3.1. *Purpose and scope of the study*

The RTE study seeks to produce scientific data of interest as part of the reflections relating to the objective of carbon neutrality by 2050 taken up by France through its successive governments[22]. RTE recalls that the French electricity production network is already highly decarbonized since it consists mainly of nuclear and hydroelectricity power plants. It should be noted that, while nuclear power represents about 70% of electricity production in 2019[23], its share constitutes slightly less than 20% of energy used by end users in France. This means that France remains heavily dependent on fossil fuels, and this production must decrease to meet the objectives of reducing greenhouse gas emissions.

The national low carbon strategy (in French: *stratégie nationale bas carbone* [SNBC]) forms the basis for climate policy within France's energy policy, and seeks to define "the procedure to be followed to lead the policy for the mitigation of greenhouse gas emissions in economically sustainable conditions in the medium and long-term"[24]. On this point, RTE relies on the roadmap set out by the SNBC, which is based on energy efficiency, low-carbon electricity and the development of uses for biomass.

21 Available at: https://www.rte-france.com/analyses-tendances-et-prospectives/bilan-previsionnel-2050-futurs-energetiques.

22 This objective has been in place since 2019 as part of the climate energy law, which sets France's objectives for climate and energy policy.

23 See the 2019 RTE electricity balance.

24 Available at: https://www.senat.fr/questions/base/2018/qSEQ181007164.html.

France produced slightly less than 500 TWh of decarbonized electricity in 2019, compared to the 645 TWh given in RTE's reference trend line (see section 5.3.2). While this gap seems small, it should be kept in mind that the average age of the reactors in operation is 36.6 years as of January 2022. The operator is currently in the investment phase (Major Refitting) with the goal of extending the operation of the current fleet beyond 40 years, up to 50 or even 60 years, which will be subject to authorization by the nuclear safety authority.

In order to fully understand the challenges of the Major Refitting program, we have constructed two bounding scenarios for the evolution of the installed nuclear capacity based on the PRIS database of reactors published by the IAEA. Only the second-generation reactors in France are considered here.

The first scenario, which we call the "40-year hypothesis", considers that each reactor will be permanently shut down after 40 years of operation starting from the first pressurized hydroelectricity test of the circuit (EHY)[25]. For reactors in operation for more than 40 years, due to the fact that the durations between two 10-year visits are slightly longer than 10 years in practice, we established a definitive shutdown on July 1, 2022. For this deliberately very unfavorable example, about 20 PWRs of 900 MWe are involved. The second scenario, called the "60-year hypothesis", is characterized by the fact that each reactor has a lifespan of 60 years.

These two scenarios are shown in Figure 5.15. The possible outcomes that lie between these two cases is represented in blue in the figure. This space is interpreted as the set of probable evolution trajectories for the second-generation reactors. We thus see that the period from 2030 to 2050 is absolutely crucial for the nuclear industry.

RTE therefore considers that the evolution of the electrical system can be carried out on the basis of two groups of trajectories:

– the "M" scenarios, characterized by a mix of 100% renewables;

– the "N" scenarios, defined by a mix of renewables and nuclear.

These two families of scenarios are based on a relatively high share of renewable energies as well as on a significant need for flexibility. Regarding economic aspects, both scenarios are characterized by high investments (renewables and/or nuclear). The scenarios are analyzed by RTE on the basis of technical, economic, environmental and societal viewpoints.

25 The dates that condition the 10-year visits of the safety authority are linked to the pressurized hydraulic testing of the circuit (French initials: EHY).

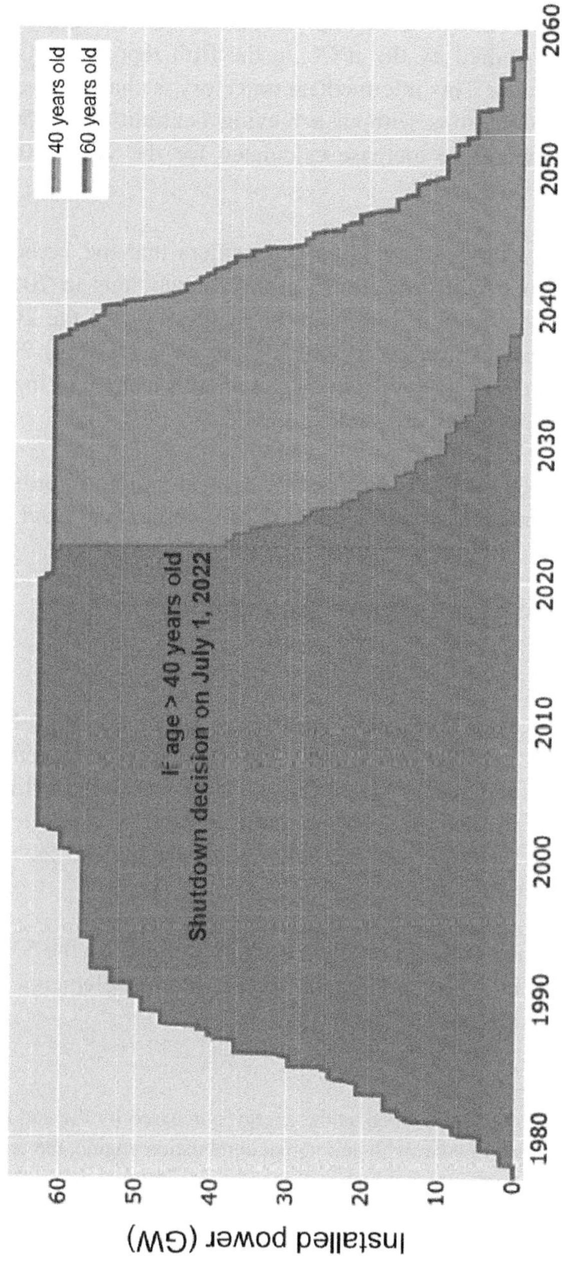

Figure 5.15. *Evolution of the nuclear installed capacity for two boundary cases (see text for the description of the two scenarios of evolution). The blue area represents a series of possibilities (source: the authors, based on IAEA/PRIS data). For a color version of this figure, see www.iste.co.uk/percebois/nuclear2.zip*

The scenarios studied by RTE are based on a complete description of the climate and its evolution, given that this evolution will have an impact on the energy production infrastructures that will be implemented in the coming decades. From among the trajectories studied by the IPCC in the fifth report (IPCC 2014), RTE used the RCP 4.5 trajectory. This intermediate trajectory is characterized by a global effort to reduce GHG emissions without achieving neutrality in 2050, or even in 2100. The average temperature increase calculated for the year 2100 is between 2.3°C and 2.9°C.

The demographic hypothesis that is used considers that the French population living in mainland France will grow to 71 million inhabitants in 2050. The GDP growth that has been chosen is in line with the suggestions of the 2016 reference scenario of the European Commission, which is about 1.3% until 2030 and then 1.7% after that. Within this framework, France maintains a share of industry of 10% of its GDP, which remains relatively modest.

Finally, RTE modeled the entire European electrical system in order to take into account the increasing integration of the network on a continental level.

5.3.2. *Consumption and production assumptions*

5.3.2.1. *Consumption*

The works done by RTE are within the framework of the 2020 National Low Carbon Strategy (SNBC), which sets final energy consumption targets[26] for France at 930 TWh in 2050, compared to the current level of 1,600 TWh. In this context, the share of electricity in the final energy consumption is expected to increase, reaching 55% of the volume of end energy consumption as compared to approximately 27% in 2020.

While RTE's approach is based on the trajectory defined by the SNBC, the data are adjusted and take into account recent socioeconomic elements (such as the COVID-19 crisis, for example).

---

26 Since the end energy amount is defined as the energy purchased by the end user, it must be used with caution. In the case of a car with an internal combustion engine, the end energy use is linked to the energy stored in the tank, which will then be transformed into mechanical energy, with at least a 70% loss. For an electric car, the end energy represents the energy of the electrical charge that is transformed almost entirely into mechanical energy. Even if the electricity is produced with fossil fuels, electric vehicles reduce total energy consumption by a factor of about 3 for the same useful energy.

The reference consumption trajectory defined by RTE therefore considers a GDP growth rate of 1.3% over the period of 2030–2050. In addition, the demographic shift has been corrected downwards.

Thus, the following three scenarios for changes in long-term consumption have been defined as:

– the reference trajectory;

– the "deep reindustrialization" scenario;

– the "sober" scenario.

Many variants of these scenarios have been defined, and the hypotheses as well as the results of the sensitivity studies are detailed in the final report.

The reference trajectory is essentially based on improving energy efficiency and replacing fossil fuels with low-carbon energy sources, mainly low-carbon electricity and biomass.

It is thus possible to show the energy supply, as well as the final energy consumption by 2050 as defined in the reference trajectory, as given in Figure 5.16.

The increase in the power from biomass seeks to reach a total potential of between 400 TWh and 450 TWh by 2050.

Biomass is intended to be consumed in the form of wood, biofuels, waste and heat, but also in the form of decarbonized gas (hydrogen, biomethane, synthetic methane, ammonia, etc.).

The reference trajectory gives an electricity supply of 645 TWh by 2050 (with consumption including network losses and hydrogen production), which represents an increase of about 35% as compared to 2020. For the reference period, this increase is broken down as follows:

– a volume effect of 47 TWh induced by population growth and economic activity;

– an "electrification of uses" effect of +324 TWh;

– an "energy efficiency" effect of –200 TWh;

– a "global warming" effect of –1 TWh.

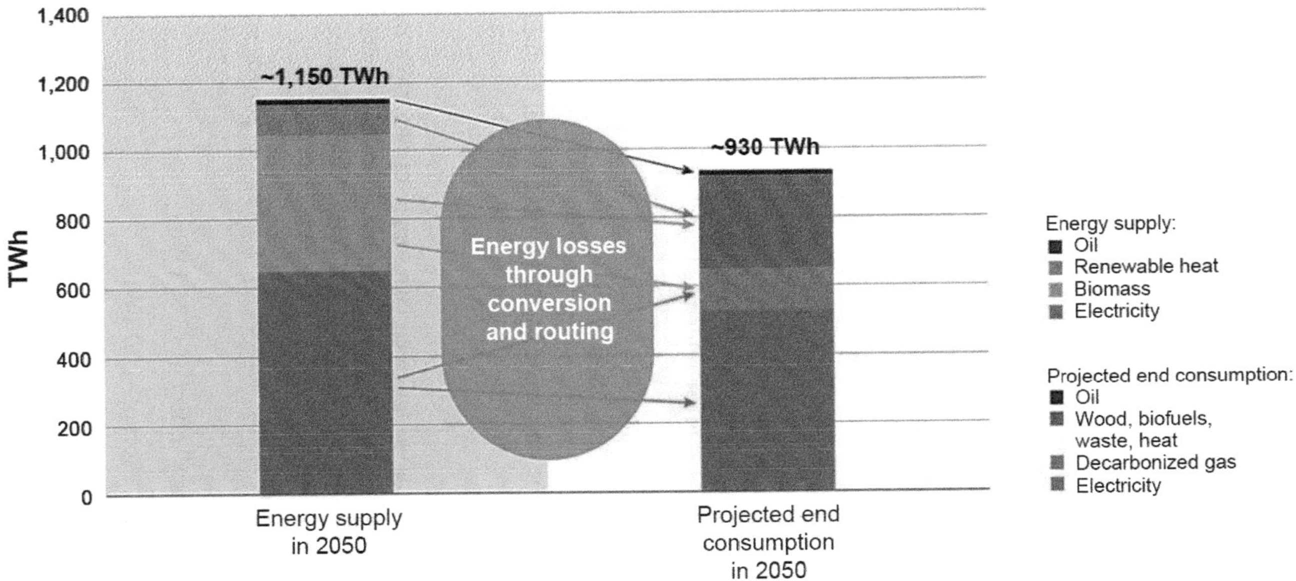

Figure 5.16. *Projection of energy supply and final energy consumption for 2050 given by the reference trajectory of the RTE study from 2021. For a color version of this figure, see www.iste.co.uk/percebois/nuclear2.zip*

RTE specifies here that the questioning of the assumptions made regarding long-term growth, lifestyle changes and the development of energy efficiency could potentially modify the assessment of future electricity demand significantly.

5.3.2.2. *Electrical production*

This section focuses on the evolution trajectories of the electrical production mix with a focus on the nuclear component of this production.

RTE recalls that French electricity production is largely decarbonized, due to the relatively large share of nuclear power in its energy mixture as well as, to a lesser extent, hydroelectric power. Since the average age of nuclear power plants is high, the power generation network will need to be renewed by 2050, which will require high levels of investment. RTE has therefore decided to structure the production assumptions into two distinct options: (1) an electrical mix composed of nuclear and renewables (scenarios N1, N2 and N03) and (2) a 100% renewable system (scenarios M1, M2 and M23), as mentioned in section 5.3.1. This choice of representation is justified by the desire to highlight the importance of political decisions in the evolution of the electrical component as part of the energy transition. Thus, the revival of a nuclear program in France, that is to say, the construction of third generation, EPR or EPR2-type generators, is political choice that will have significant structural effects, which justifies the line of demarcation of the scenarios studied.

The scenarios studied were constructed from a detailed examination of the nuclear component. In any case, the capacity of the existing nuclear network is likely to decrease in the next three decades. As we have shown in Figure 5.17, there are many possible ways to "smooth the edges" in order to avoid a "cliff effect". The shutdown of 12 reactors is planned for the period 2020–2035[27] (as per the 2018 PPE, which the 2023 PPE may or may not keep in place). Additionally, the shutdowns will be more frequent, and there will be no more second-generation reactors in operation by 2060. Naturally, the success of the Major Refitting program is a key assumption for the trajectories studied here.

Regarding the construction of new reactors, RTE's "N" scenarios are based on the New Nuclear France (NNF) program, the objective of which is to build three pairs of EPR2s, or six reactors in total, between 2035 and 2045 for an installed capacity in the order of 10 GW.

---

27 In addition to the two Fessenheim reactors shut down in February and June 2020, respectively.

The trajectory of scenario N1 is based on a rate of 5 years for the construction and commissioning of a pair of EPR2 reactors starting in 2045. In 2060, this trajectory therefore includes six pairs of EPR2s, in addition to the EPRat Flamanville, with an installed capacity in the order of 20 GW in 2060. The trajectories of scenarios N2 and N03 are more favorable to nuclear power and assume the commissioning of a pair of EPR2s every 3 years starting in 2044, reaching an installed capacity of 40 GW by 2060.

SMRs have been included in the RTE study via the "N03" scenario, which takes into account the potential development of 4GW of installed power, in addition to the EPR2s, by 2050. On the other hand, the fourth-generation reactors have not been included into the study. Indeed, as the ASTRID project (Sodium FNRs) has been postponed to the end of the century, the PPE does not foresee any possible implementation before the second half of the 21st century.

Figures 5.17(A–C) summarize the nuclear trajectories in the six scenarios studied by RTE.

The evolution of electricity production in the scenarios studied by RTE is based in all cases on a strong growth in renewable energies, in line with many global projections, in particular those carried out by the international energy agency. The relative shares of the different renewable production sources depend on the scenarios. However, the minimum value of the total electrical output is 50%. More specifically, RTE assumes a moderate level of development of hydroelectric power capacities. The growth of run-of-river and lake-based hydro power, though less than the growth by pumped storage power stations (PSPS) will be more significant, adding up to an additional 3 GW between 2020 and 2050. PV solar generation also has strong growth assumptions, with installed capacities ranging between 70 GW and 215 GW by 2050. Regarding onshore wind, the deployment assumptions have found a certain degree of support and vary between 43 GW and 75 GW of installed power by 2050. It is considered that there are strong growth prospects for offshore wind. The deployment assumptions range from 22 GW up to 62 GW of installed power for the year 2050.

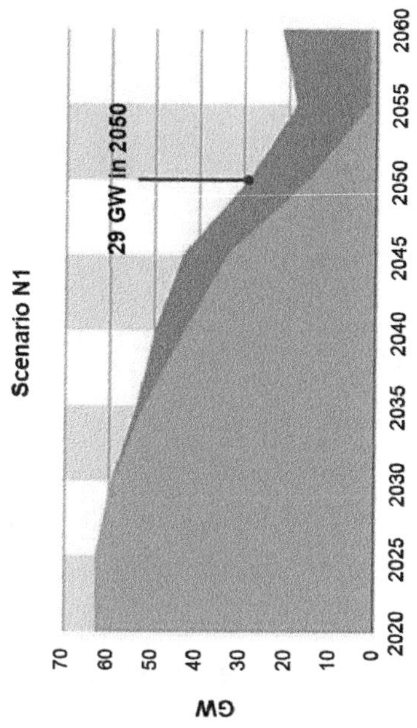
Scenario N1
29 GW in 2050
GW
70
60
50
40
30
20
10
0
2020
2025
2030
2035
2040
2045
2050
2055
2060

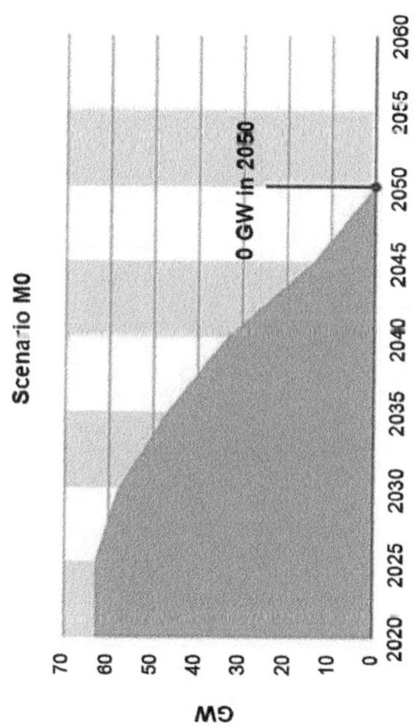
Scenario M0
0 GW in 2050
GW
70
60
50
40
30
20
10
0
2020
2025
2030
2035
2040
2045
2050
2055
2060

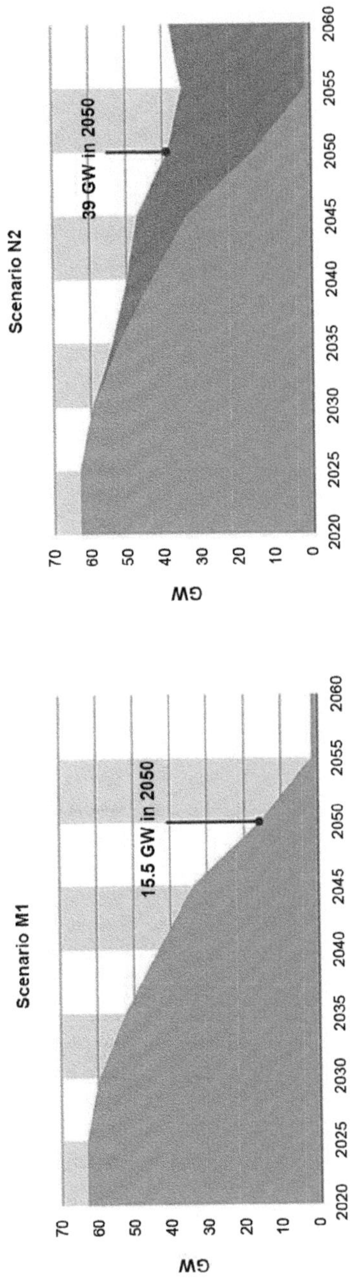
Scenario N2
39 GW in 2050
GW
70
60
50
40
30
20
10
0
2020
2025
2030
2035
2040
2045
2050
2055
2060
Scenario M1
15.5 GW in 2050
GW
70
60
50
40
30
20
10
0
2020
2025
2030
2035
2040
2045
2050
2055
2060

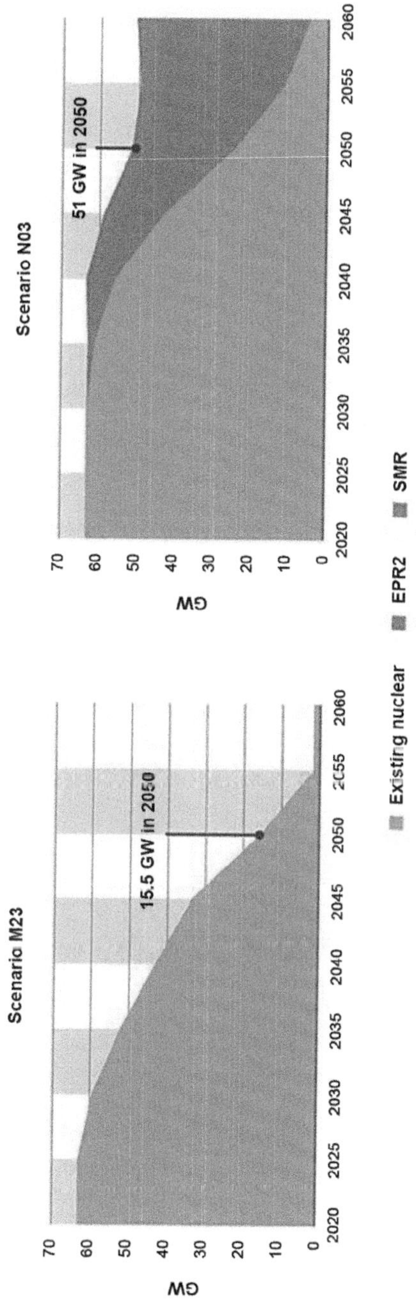

Figure 5.17. *Synthesis of the evolution trajectories of nuclear power by type of technology in the six RTE study scenarios (figure taken from the 2021 report). For a color version of this figure, see www.iste.co.uk/percebois/nuclear2.zip*

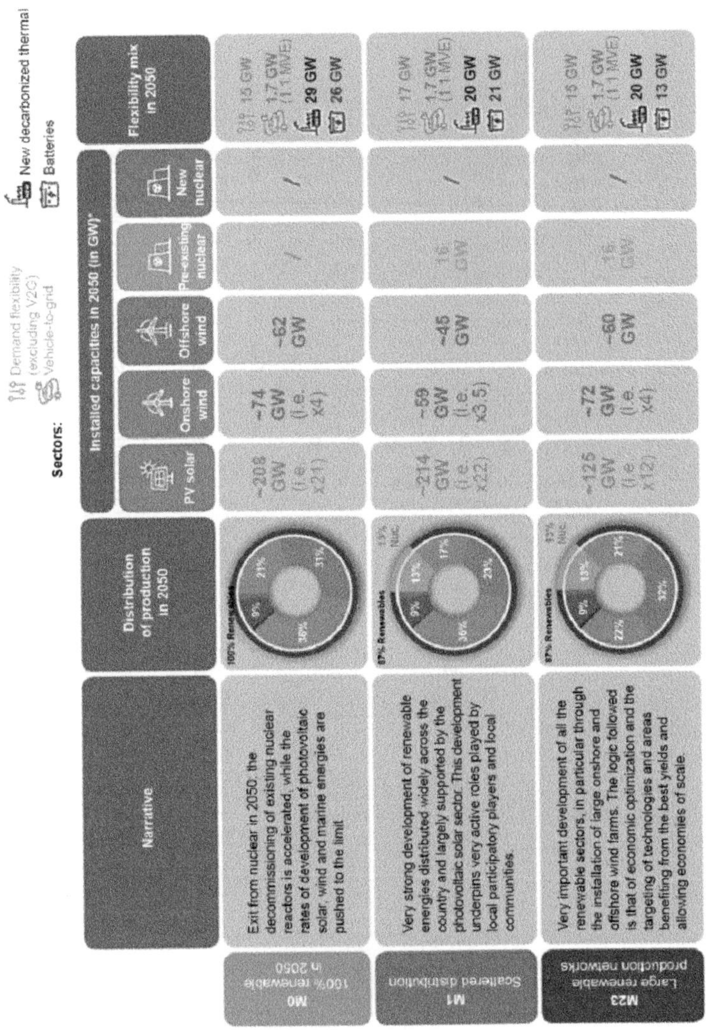
Sectors:
Demand flexibility (excluding V2G)
Vehicle-to-grid
New decarbonized thermal
Batteries
Narrative
Distribution of production in 2050
Installed capacities in 2050 (in GW)*
PV solar
Onshore wind
Offshore wind
Pre-existing nuclear
New nuclear
Flexibility mix in 2050
M0
100% renewable in 2050
Exit from nuclear in 2050: the decommissioning of existing nuclear reactors is accelerated, while the rates of development of photovoltaic solar, wind and marine energies are pushed to the limit
~208 GW (i.e. x21)
~74 GW (i.e. x4)
~62 GW
/
/
15 GW
1.7 GW (1.1 MVE)
29 GW
26 GW
M1
Scattered distribution
Very strong development of renewable energies distributed widely across the country and largely supported by the photovoltaic solar sector. This development underpins very active roles played by local participatory players and local communities.
87% Renewables
13% Nuc.
~214 GW (i.e. x22)
~59 GW (i.e. x3.5)
~45 GW
16 GW
/
17 GW
1.7 GW (1.1 MVE)
20 GW
21 GW
M23
Large renewable production networks
Very important development of all the renewable sectors, in particular through the installation of large onshore and offshore wind farms. The logic followed is that of economic optimization and the targeting of technologies and areas benefiting from the best yields and allowing economies of scale.
87% Renewables
13% Nuc.
~125 GW (i.e. x12)
~72 GW (i.e. x4)
~60 GW
16 GW
/
15 GW
1.7 GW (1.1 MVE)
20 GW
13 GW

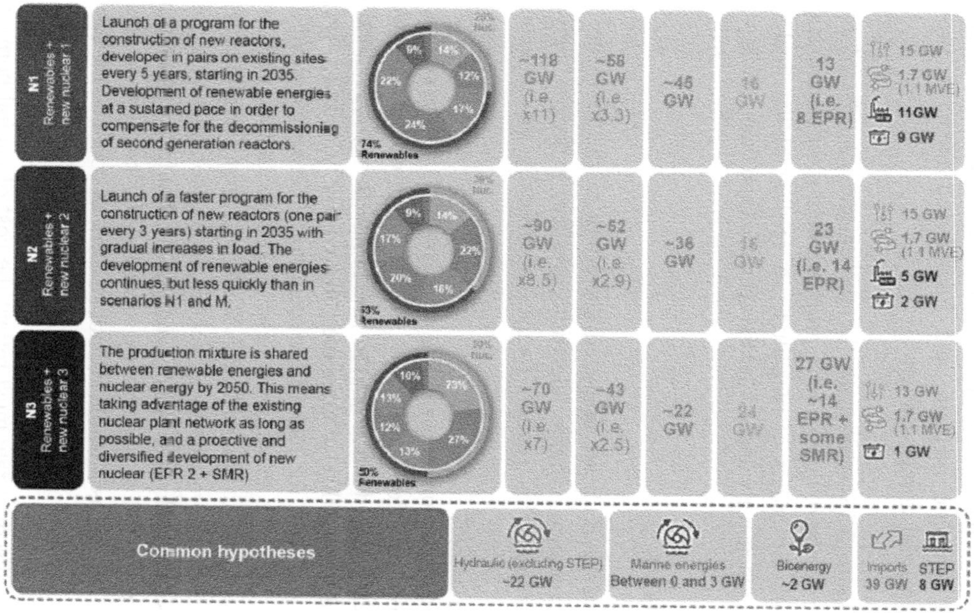

*The amounts and shares of energy are expressed in relation to the reference consumption scenario.

Figure 5.18. *Synthesis of production mix scenarios for 2050 created by RTE from the 2021 study. For a color version of this figure, see www.iste.co.uk/percebois/ nuclear2.zip*

In all the scenarios, the high proportion of intermittent renewable energy makes it necessary to have the capability to use controllable production units in order to ensure the supply is secure at all times. Thermal power plants are perfectly adapted, but under their current operations based on fossil fuels, they do not allow them to be in line with the strategy for reducing greenhouse gas emissions. Therefore, the reduced-carbon gaseous fuels projected to be used in these plants are low-carbon hydrogen, bio-methane, synthetic methane and fossil-based methane, provided together with a CO_2 capture and storage system.

The two parts of Figure 5.18 summarize the six trajectories selected for the electrical production. The summary that we have proposed here is far from exhaustive but allows us to establish the orders of magnitude. A reading of the final report remains essential to comprehensively understand the scenarios and their results.

5.3.3. *Economic analysis*

The scenarios created by RTE are analyzed according to the following main topics: Europe, supply security, climate, electricity system, hydrogen, networks, environment and economics. This section focuses specifically on an economic analysis.

The RTE methodology relies on the complete calculation of costs to compare the different energy transition scenarios. In this way, the RTE approach creates two key indicators: the investment expenditure over the period and the updated total cost. RTE.s approach is original in that it integrates the cost of flexibility into the calculation of the estimated costs for each of the scenarios. The cost of flexibility is in fact the *system* cost, that is to say, the cost linked to factors such as the intermittency of renewables (see Chapter 3 of Volume 1 and Chapter 2 of this volume). It is thus necessary to provide for a back-up, or for storage (in the form of power-to-gas) to address the lack of wind or sunlight. It should be noted that the scope of this study is restricted to the energy system. The positive or negative externalities induced for society as a whole are not studied, as could be the case in a socioeconomic analysis.

RTE justifies that the discount rate must reflect the cost of capital borne by the technologies which are deployed. This approach differs from the approach using the socio-economic discount rate chosen as part of the public economic calculation (see Chapter 2 of Volume 1). RTE bases its findings on the assumption of a uniform cost of capital for all technologies, set at 4% with variants, meaning that this rate can vary between 1% and 7%.

For second-generation nuclear plants, the cost of extending the lifespan (below 60 years) is set at €650/kW and €440/kW for before and after 2035, respectively. The fixed and variable operating costs are set at €23/MWh and include, among others, the costs of managing spent fuels and nuclear waste. RTE has also included the cost of a nuclear accident at the operator level by taking into account the civil liability insurance contribution which ensures coverage of up to 700 million euros (see Chapter 4), with the rest being covered by the government, and therefore outside the scope of the study.

Regarding the evolution of the costs of the new nuclear power plant, the economic data are characterized by a relatively high level of uncertainty. A learning effect has been integrated into the findings, leading to a trend decrease in investment costs between 2020 and 2045 of up to €4,500/kW. These data are subject to sensitivity studies. The cost of construction for SMRs was set at €5,500/kW, with a fairly low level of confidence, since the uncertainties as to the costs of this type of reactor are high. The financing conditions for the new nuclear power plant, as well as the effective load factors, which will strongly depend on the electricity mix, will be likely to significantly influence the economic data relating to nuclear power. The dismantling costs are taken into account through a provision made during the commissioning of the reactor amounting to €180/kW. The downstream effects of the nuclear power cycle have also been integrated into the economic calculation based on an estimate classified as "conservative". The amounts set for treatment/recycling are between €3/MWh and €7/MWh and depend on the size of the nuclear plant network. The costs of nuclear waste management are based on a high assessment of the costs of the Cigéo project, at €31 billion$_{2020}$within the scope of the study, with an increase of €10 billion$_{2020}$ which integrates the potential additional spent fuels (spent MOX, URE and UOX), which are not intended today to be stored at Cigéo. The costs related to storage in the deep geological layer have therefore been estimated at around €2.1/MWh.

The cost assumptions relating to other means of electrical production are also detailed in the report. Prospects for a sharp decrease in the costs of wind and PV solar production are projected in accordance with the data from the specialized literature. The analysis of the production costs of the main sectors shows that renewable energies are generally cheaper than new nuclear sources, if the network costs and flexibility are not taken into account. However, the orders of magnitude are quite comparable with uncertainties about the costs, which do not allow for any indisputable conclusions.

RTE highlights the relevance of continuing the operation of existing nuclear reactors over the medium term, that is, through 2030–2035. In the longer term (up to 2060), two important results from the RTE study can be identified. First, the results concerning investment expenditures (see Figure 5.19) show that the orders of magnitude are comparable, though more favorable than scenarios with nuclear power ("N" scenarios). The fundamental differences appear at the level of the renewables and nuclear, but also in terms of flexibility.

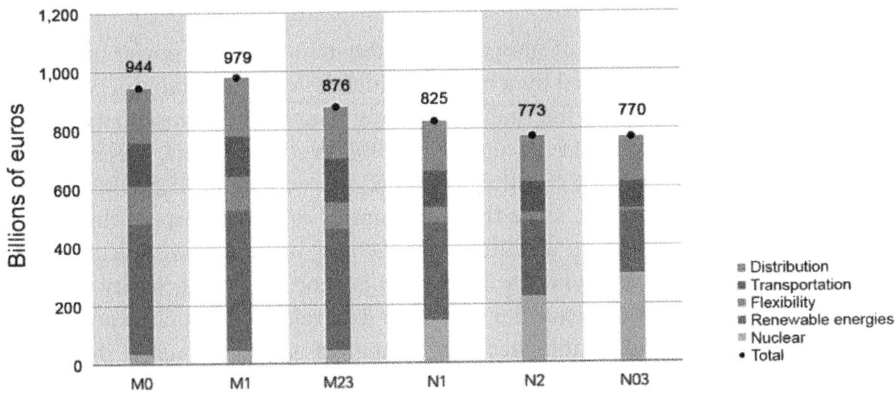

Figure 5.19. *Cumulative capital expenditure over the period of 2020–2060 for the electrical system. Curve from the RTE report (2021). For a color version of this figure, see www.iste.co.uk/percebois/nuclear2.zip*

The total cost of the electrical system tends to increase for all the trajectories considered. The median by 2060 is +15% compared to 2020. The result of the calculation concerning the full cost in 2060 for the six trajectories studied is presented in Figure 5.20.

The total costs represented take into account the production costs as well as the flexibility and network costs. The scenarios incorporating new nuclear power in 2050 are slightly cheaper than scenarios without the construction of new reactors. The explanation lies in the fact that the needs for flexibility for scenarios with a high deployment of renewables compensate for the lower production cost for these means of production. Here, however, RTE insists on the fact that the differences are too small to allow for a clear conclusion to be made and recommends that additional non-economic indicators be taken into account.

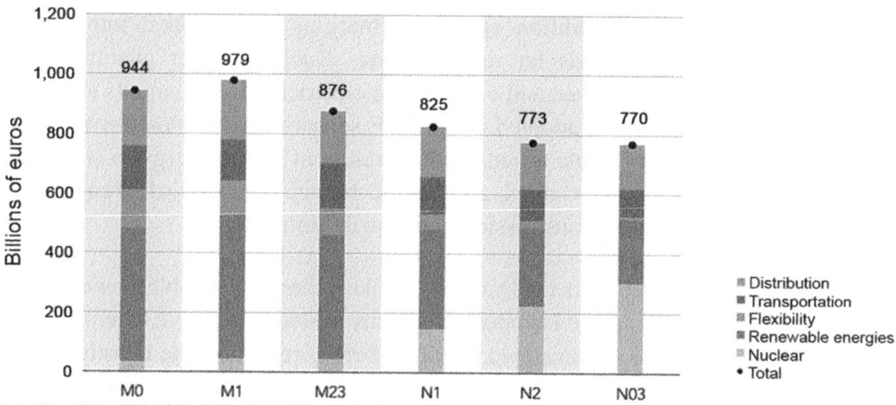

Figure 5.20. *Annualized full scenario costs for 2060. Curve from the RTE report (2021). For a color version of this figure, see www.iste.co.uk/percebois/nuclear2.zip*

5.3.4. *Comments in response to the RTE study*

Several criticisms have been made about this study. The first concerns the assumption of a 40% decrease in end energy demand between 2019 and 2050, considered very optimistic by some. The second concerns the estimated electricity demand of 645 TWh in 2050. For some, this is too low a figure in a context where the electrification of uses will become widespread. In fact, RTE studies two variants, at 554 and 752 TWh, respectively. In 2021, the Academy of Technologies estimated the level needed for 2050 at a minimum of 730 TWh of electricity. The Academy of Sciences, for its part, estimated that the electricity demand could even reach around 900 TWh by that time. The Negawatt association thinks it would be possible to limit the demand for electricity to 272 TWh, but given the assumptions made about energy efficiency, this would mean a society with more austere consumption habits, very different from current society.

The demand for electricity by 2050 in France could be between 730 and 840 TWh according to the Academy of Technologies, and this will depend in particular on the share reserved for nuclear power and the hypotheses for the development of hydrogen. The demand for electricity will be made even greater as the share of intermittent renewables will be greater, insofar as it will then be necessary to store a high proportion of this electricity in the form of hydrogen or methane. Another essential parameter of any scenario is the additional power put into service at peak times; this regularly exceeds 90 GW and has even reached 102 GW in 2012. The Academy estimates that peak power could reach or even exceed 130 GW in 2050. In addition, significant margins must be taken into account when considering time periods far in the future. Nuclear power guarantees the continued supply of decarbonized and controllable electricity. Increasing its share (i.e. to above 50%, the assumption adopted in the RTE scenario that is most favorable for nuclear power) makes it possible to reduce the amount of stored energy to counteract the intermittent nature of renewables, and also to reduce the costs of strengthening and extending the electricity transmission and distribution network.

It should also be noted that nuclear power, like other controllable power plants, contributes to the stability of the European electricity system, a positive externality that is rarely noted, much less remunerated. Waiting before restarting the construction of new reactors is not an option, the Academy recalls, given the inevitable disappearance of skills and nuclear manufacturing resources in the absence of any new projects.

Table 5.3 summarizes the conclusions of the study carried out by the Academy of Technologies in its 50% nuclear scenario in 2050 (in France).

	2019 Consumption (TWh)	2050 Consumption (TWh)
Electricity consumption	470	840
(of which are produced outside biomass)	(460)	(800)
(of which are electricity from biomass)	(10)	(40)
Biomass heat	178	325
Fossil fuels	1,460 (of which 919 are oil, 455 gas, and 86 coal)	0
Total	2,108	1,165

Table 5.3. *Results of the study by the Academy of Technologies from 2021*

Taking into account the loss on the electricity transmission and distribution networks (about 7%), 900 TWh of electricity would have to be produced, of which 450 TWh would be nuclear power (50%), which requires an installed level of nuclear power of 68 GW. With an estimated hydroelectricity production of 60 TWh and a biomass electricity production of 40 TWh, this would result in a controllable electricity production in the order of 550 TWh, versus a non-controllable electricity production (wind and PV solar) in the order of 350 TWh. It can be seen that the necessary nuclear power (68 GW) is significantly higher than the figure from the strong nuclear scenario given by the RTE (50% nuclear, 51 GW). This is largely explained by the assumption of a greater total energy consumption than that used in the RTE scenarios.

5.4. Conclusion

This chapter, which focuses on the role of nuclear power in energy scenarios, demonstrates the way in which public decision-makers have historically integrated atomic energy into their policies to confront various crises (involving oil, geopolitics and the climate). As the complexity of technological and economic scenarios requires that the acceptability of policies and the breaking points in the trajectories of production and consumption must be included, it becomes difficult to defend the best solutions given by models, both for new technologies and for the nuclear policy to be followed. The role of nuclear power in prospective scenarios thus appears as the assumption of a starting point rather than an outcome, given that it is highly dependent on policies and societies, which in some global-level scenarios reaches up to 800 GW in 2050 (IEA 2021). The decarbonization of the world's economies, which has become the goal of a significant number of countries, will require the creation of a significant amount of intermittent renewable energies, which raises the question of the flexibility of electrical systems and the adaptation of nuclear power to the new context of sub-hourly variability. The solution offered by SMRs may serve as a complementary and innovative power source thanks to its flexibility and ability to be located on isolated sites.

The scenarios for anticipating flexibility needs allow for the participation of nuclear reactors, taking into account their technical constraints such as adjustment speed, the minimum safety threshold of the reactors and the share of the nuclear network operating in base or load monitoring mode. In the case of France, it has been shown that a combined mix of 25% nuclear and 71% renewables in the electricity supply by 2050 requires more deep cycles operated by the reactors than the current budget allows, which means it will be necessary to have reactors that are faster or to have their licenses reviewed to avoid expensive refittings or the early decommissioning of the reactors. On the electricity market, this impacts gas-based production as well as the supply of renewable energy or demand, and redefines investment and infrastructure choices on the basis of three factors underlying load monitoring operations: the management of uranium when a reactor under-utilizes production capacity; the ability of the nuclear network to meet demand peaks when the political choice is made based on energy and not power; and the inclusion of flexibility in the segments of the decentralized electricity market in a context of the centralized management of the nuclear network. Alternatives to the provision of nuclear flexibility would facilitate the integration of renewables, such as the production of hydrogen with surplus nuclear power, and/or the deployment of SMR reactors, as proposed by one of the RTE, called N03.

The "Energy Futures 2050" study by RTE (2021) creates a reference trajectory for the decarbonization of the economy in France, based on an increase in electricity consumption and on two families of scenarios for the electrical production mix, assigning a leading role to renewables (M) or nuclear (N). This means that the need for flexibility does exist, which presupposes energy storage solutions in the form of hydroelectricity pumping (3 GW of additional STEP) and of the power-to-gas type, as well as the construction of SMRs (4 GW). By breaking down the trajectory of the possible decommissioning of nuclear reactors, on the basis of an average reactor age of 36.6 years, we note that the period from 2030 to 2050 is crucial for the nuclear industry, and that only a smoothing of the reactor shutdowns can make it possible to avoid the cliff effect and periods of disruption of energy supply; the revival of a nuclear program for the construction of EPR-type reactors would thus be essential. Since the investment expenditures in renewables and nuclear power are somewhat comparable – although favorable to scenarios with nuclear power – the fundamental differences between the scenarios can be found in terms of the flexibility, storage and extension of the electricity network. Since a clear demarcation of the scenarios has not been put forward, additional indicators will have to be taken into account when making the political and societal choice of deciding on the future electricity mix, such as the acceptability of wind farms and large-scale solar power plants, the technological and geographical diversification of the electricity mix, and industrial and climate-related considerations.

As decentralization is one of the prerequisites of the citizen energy transition in order to involve all players involved, including consumers, the role of territorial entities in making decisions is clearly increasing, but the choice of nuclear power is above all a choice of industrial planning by the federal government, as it is based on a long-term commitment, involving multiple dimensions and participants with a strong financial investment capacity. Two different decision-making levels are superimposed on each other, and their complementary nature comes in addition to that of the nature of the inputs of each method of electricity production, renewable and nuclear. In Europe, the European Commission has announced that it is committed to its choice in favor of nuclear power, providing 500 billion euros of investment in the construction of new power plants as part of a new taxonomy that classifies nuclear power as green energy. This label seeks to facilitate the financing of installations contributing to the fight against climate change and makes it possible to benefit from reductions in financing costs for both nuclear and gas power plants under specific eligibility criteria: transitional energy, time-limited, guaranteeing waste management, adopt the best available technologies and replace more polluting energy sources, such as coal. According to the outlook of the EU, in which we have less than 30 years left to decarbonize the economy, renewable energies remain a priority, but are not sufficient to meet the growing demand for electricity on their own. Ultimately, in addition to the context of the climate emergency, the high level of countries' energy dependence must also be considered. This makes nuclear policy, as a decarbonized and controllable solution, an option – albeit an insufficient one – for achieving sustainability for the following decades, with the objective of limiting global warming to 2°C by 2100.

5.5. References

ADEME (2016). A 100% renewable electricity mix? Analysis and optimisation. Final report, Agence de l'Environnement et de la Maîtrise de l'Énergie, Paris, France.

ADEME (2021). Transition(s) 2050. Choisir maintenant, agir pour le climat.

Alimou, Y., Maïzi, N., Bourmaud, J.-Y., Li, M. (2020). Assessing the security of electricity supply through multi-scale modeling: The TIMES-ANTARES linking approach. *Applied Energy*, 279, 115717.

Allard, S., Mima, S., Debusschere, V., Quoc, T.T., Criqui, P., Hadjsaid, N. (2020). European transmission grid expansion as a flexibility option in a scenario of large scale variable renewable energies integration. *Energy Economics*, 87, 104733.

ANCRE (2013). Scénarios de l'ANCRE pour la transition énergétique.

ANL (2005). Small modular fast reactors design description. Argonne National Laboratory.

ANL (2018). Balancing nuclear and renewable energy. Argonne National Laboratory.

AREVA-EDF (2009). Pre-construction safety report, sub chapter 1.2, general description of the unit – UK EPR Issue 01.

ASN (2017). PNGMDR 2016-2018. ASN.

Bauer, N., Calvin, K., Emmerling, J., Fricko, O., Fujimori, S., Hilaire, J., Eom, J., Krey, V., Kriegler, E., Mouratiadou, I. et al. (2017). Shared socio-economic pathways of the energy sector-quantifying the narratives. *Global Environmental Change*, 42, 316–330.

Bertsch, J., Growitsch, C., Lorenczik, S., Nagl, S. (2018). Flexibility in Europe's power sector – An additional requirement or an automatic complement? *Energy Economics*, 53, 118–131.

Bruynooghe, C., Eriksson, A., Fulli, G. (2010). Load-following operating mode at nuclear power plants (NPPs) and incidence on operation and maintenance (O&M) costs. Compatibility with wind power variability. Publications Office of the European Union.

Cany, C. (2017). Interactions entre énergie nucléaire et énergies renouvelables variables dans la transition énergétique en France ; adaptations du parc électrique vers plus de flexibilité. PhD Thesis, Université Paris-Saclay.

Cany, C., Mansilla, C., Mathonnière, G., Da Costa, P. (2017). Adapting the French nuclear fleet to integrate variable renewable energies via the production of hydrogen: Towards massive production of low carbon hydrogen? *International Journal of Hydrogen Energy*, 42(19), 13339–13356.

Cany, C., Mansilla, C., Mathonnière, G., Da Costa, P. (2018a). Nuclear power supply: Going against the misconceptions. Evidence of nuclear flexibility from the French experience. *Energy*, 151, 289–296.

Cany, C., Mansilla, C., Mathonnière, G., Da Costa, P. (2018b). Nuclear contribution to the penetration of variable renewable energy sources in a French decarbonised power mix. *Energy*, 150, 544–555.

Capros, P., Mantzos, L., Criqui, P., Kouvaritakis, N., Soria Ramirez, A., Schrattenholzer, L., Vouyoukas, E.L. (1999). *Climate Technology Strategies 1: Controlling Greenhouse Gases. Policy and Technology Options*. Physica-Verlag, Heidelberg.

Cour des comptes (2014). Le coût de production de l'électricité nucléaire.

Cour des comptes (2018). L'aval du cycle du combustible nucléaire.

CRE (2018). Étude sur les perspectives stratégiques de l'énergie. E-cube report.

Criqui, P. and Mima, S. (2012). European climate–energy security nexus: A model based scenario analysis. *Energy Policy*, 41, 827–842.

Criqui, P., Mima, S., Menanteau, P., Kitous, A. (2015). Mitigation strategies and energy technology learning: An assessment with the POLES model. *Technological Forecasting and Social Change*, 90, 119–136.

Després, J., Mima, S., Kitous, A., Criqui, P., Hadjsaid, N., Noirot, I. (2017). Storage as a flexibility option in power systems with high shares of variable renewable energy sources: A POLES-based analysis. *Energy Economics*, 64, 638–650.

Edenhofer, O., Knopf, B., Barker, T., Baumstark, L., Bellevrat, E., Chateau, B., Criqui, P., Isaac, M., Kitous, A., Kypreos, S. et al. (2010). The economics of low stabilization: Model comparison of mitigation strategies and costs. *The Energy Journal*, 31, 11–48.

EUR (2012). European utility requirements for LWR nuclear power plant. Volume 2: Generic nuclear island requirements [Online]. Available at: https://www.europeanutilityrequirements.eu/EURdocument.aspx.

Feutry, S. (2013). Load following EDF experience feedback. Paper presented at IAEA Technical Meeting on Flexible (Non-baseload) Operation Approaches for Nuclear Power Plants, Paris.

Frisch, J.R. (1983). *Energy 2000-2020: World Prospects and Regional Stresses*. Graham & Trotman Ltd. and Springer, Dordrecht.

Goldemberg, J., Johansson, T.B., Reddy, A.K., Williams, R.H. (1988). *Energy for a Sustainable World*. John Wiley & Sons, New York.

Göransson, L. and Johnsson, F. (2009). Dispatch modeling of a regional power generation system – Integrating wind power. *Renewable Energy*, 34(4), 1040–1049.

Goutte, S. and Vassilopoulos, P. (2019). The value of flexibility in power markets. *Energy Policy*, 125, 347–357.

Grübler, A. (2010). The costs of the French nuclear scale-up: A case of negative learning by doing. *Energy Policy*, 38(9), 5174–5188.

Grübler, A., Jefferson, M., Nakićenović, N. (1996). Global energy perspectives: A summary of the joint study by the International Institute for Applied Systems Analysis and World Energy Council. *Technological Forecasting and Social Change*, 51(3), 237–264.

Grübler, A., Wilson, C., Bento, N., Boza-Kiss, B., Krey, V., McCollum, D.L., Rao, N.D., Riahi, K., Rogelj, J., De Stercke, S. et al. (2018). A low energy demand scenario for meeting the 1.5 °C target and sustainable development goals without negative emission technologies. *Nature Energy*, 3(6), 515–527 [Online]. Available at: https://doi.org/10.1038/s41560-018-0172-6.

Häfele, W., Anderer, J., McDonald, A., Nakicenovic, N. (1981). *Energy in a Finite World: Paths to a Sustainable Future, Volume 1*. Ballinger, Cambridge, MA.

IAEA (2018). Non-baseload operations in nuclear power plants: Load-following and frequency control flexible operations. Report NP-T-3.23.

International Energy Agency (2021). Net zero by 2050. A roadmap for the global energy sector.

Jenkins, J., Zhou, Z., Ponciroli, R., Vilim, R.B., Ganda, F., de Sisternes, F., Botterud, A. (2018). The benefits of nuclear flexibility in power system operations with renewable energy. *Applied Energy*, 222, 872–884.

Johnstone, N. and Haščič, I. (2012). Increasing the penetration of intermittent renewable energy: Innovation in energy storage and grid management. Energy and Climate Policy: Bending the Technological Trajectory.

Kouvaritakis, N., Soria, A., Isoard, S. (2000). Modelling energy technology dynamics: Methodology for adaptive expectations models with learning by doing and learning by searching. *International Journal of Global Energy Issues*, 14(1/4), 104–115.

Kriegler, E., Weyant, J.P., Blanford, G.J., Krey, V., Clarke, L., Edmonds, J., Fawcett, A., Luderer, G., Riahi, K., Richels, R. et al. (2014). The role of technology for achieving climate policy objectives: Overview of the EMF 27 study on global technology and climate policy strategy. *Climatic Change*, 123, 353–367.

Kriegler, E., Petermann, N., Krey, V., Schwanitz, V.J., Luderer, G., Ashina, S., Bosetti, V., Eom, J., Kitous, A., Méjean, A. et al. (2015). Diagnostic indicators for integrated assessment models of climate policy. *Technological Forecasting and Social Change*, 90, 45–61.

Leurent, M., Da Costa, P., Rämä, M., Persson, U., Jasserand, F. (2018). Cost-benefit analysis of district heating systems using heat from nuclear plants in seven European countries. *Energy*, 149, 454–472.

Loisel, R., Alexeeva, V., Zucker, A., Shropshire, D. (2018). Load-following with nuclear power: Market effects and welfare implications. *Progress in Nuclear Energy*, 109, 280–292.

Ludwig, H., Salnikova, T., Stockman, A., Waas, U. (2010). Load cycling capabilities of German nuclear power plants. *International Journal for Nuclear Power*, 55(8).

Mantripragada, H. and Rubin, E. (2018). Techno-economic analysis methods for nuclear power plants (June 18, 2018). USAEE Working Paper No. 18-346 [Online]. Available at: https://ssrn.com/abstract=3198421.

Meadows, D.H., Goldsmith, E.I., Meadow, P. (1972). *The Limits to Growth*, Volume 381. Earth Island Limited, London.

Ministère de la Transition écologique et solidaire (2015). Programmation pluriannuelle de l'énergie. Ministère de la Transition écologique et solidaire.

Morilhat, P., Feutry, S., Lemaitre, C., Favennec, J.M. (2019). Nuclear power plant flexibility at EDF. *VGB PowerTech*, 99(5), 32–41.

Muniglia, M. (2017). Optimisation du pilotage d'un réacteur à eau pressurisée dans le cadre de la transition énergétique à l'aide d'algorithmes évolutionnaires. PhD Thesis, Université Paris Saclay.

Nakicenovic, N., Alcamo, J., Grübler, A., Riahi, K., Roehrl, R.A., Rogner, H.-H., Victor, N. (2000). Special report on emissions scenarios (SRES). A Special Report of Working Group III of the Intergovernmental Panel on Climate Change. Cambridge University Press.

Negawatt Association (2017). Scénario Negawatt 2017-2050. Report 170126a.

Negawatt Association (2022). La transition énergétique au cœur d'une transition sociétale.

Newell, R., Raimi, D., Aldana, G. (2019). Global energy outlook 2019: The next generation of energy. *Resources for the Future*, 8–19.

Nicolosi, M. (2010). Wind power integration and power system flexibility – An empirical analysis of extreme events in Germany under the new negative price regime. *Energy Policy*, 38, 7257–7268.

Nordhaus, W. (2019). Can we control carbon dioxide? (From 1975). *American Economic Review*, 109(6), 2015–2035.

OECD/NEA (2011). Technical and economic aspects of load following with nuclear power plants. OECD/NEA, Paris, France.

OECD/NEA (2012). Nuclear energy and renewables: System effects in low-carbon electricity systems. OECD/NEA, Paris, France.

OECD/NEA (2015). Projected costs of generating electricity. OECD/NEA, Paris, France.

OECD/NEA (2019). The costs of decarbonisation: System costs with high shares of nuclear and renewables. OECD/NEA, Paris, France.

O'Neill, B.C., Kriegler, E., Ebi, K.L., Kemp-Benedict, E., Riahi, K., Rothman, D.S., van Ruijven, B.J., van Vuuren, D.P., Birkmann, J., Kok, K. et al. (2017). The roads ahead: Narratives for shared socioeconomic pathways describing world futures in the 21st century. *Global Environmental Change*, 42, 169–180 [Online]. Available at: https://doi.org/10.1016/j.gloenvcha.2015.01.004.

Persson, J., Andgren, K., Henriksson, H., Loberg, J., Malm, C., Pettersson, L., Sandström, J., Sigfrids, T. (2012). Additional costs for load-following nuclear power plants experiences from Swedish, Finnish, German, and French nuclear power plants. Elforsk report 12:71 [Online]. Available at: https://energiforskmedia.blob.core.windows.net/media/21094/additional-costs-for-load-following-nuclear-power-plants-elforskrapport-12-71.pdf.

Pezzey, J.C.V. and Toman, M.A. (2002). The economics of sustainability: A review of journal articles. Discussion Paper 02-03, Resources for the Future, Washington, USA [Online]. Available at: https://media.rff.org/documents/RFF-DP-02-03.pdf.

Pindyck, R.S. (2013). Climate change policy: What do the models tell us? *Journal of Economic Literature*, 51(3), 860–872.

Poncelet, K., Delarue, E.., Duerinck, J., Six, D., D'Haeseleer, W. (2014). The importance of integrating the variability of renewables in long-term energy planning models. In *IAEE European Energy Conference*, Rome.

Ponciroli, R., Wang, Y., Zhou, Z., Botterud, A., Jenkins, J., Vilim, R.B., Ganda, F. (2017). Profitability evaluation of load-following nuclear units with physics-induced operational constraints. *Nuclear Technology*, 200(3), 189–207.

Quinet, A. and Bueb, J. (2019). La valeur de l'action pour le climat. *France Stratégie*.

RTE (2017). Bilans prévisionnels sur l'évolution de l'équilibre offre/demande d'électricité.

RTE (2018). Production réalisée par groupe.

RTE (2021). Conditions et prérequis en matière de faisabilité technique pour un système électrique avec une forte proportion d'énergies renouvelables à l'horizon 2050.

Sachs, J., Guerin, E., Mas, C., Schmidt-Traub, G., Tubiana, L., Waisman, H., Colombier, M. (2014). Pathways to deep decarbonization-2014 report [Online]. Available at : https://www.osti.gov/etdeweb/biblio/22328772.

Scamman, D. and Newborough, M. (2016). Using surplus nuclear power for hydrogen mobility and power-to-gas in France. *International Journal of Hydrogen Energy*, 41(24), 10080–10089.

Shift Project (2018). Future studies on the energy transition. Technical file #1 – Draft version. Information and recommendations for scenario producers [Online]. Available at: https://theshiftproject.org/wp-content/uploads/2019/04/Rapport-intermédiaire-Power-Systems-2050.pdf.

Schrattenholzer, L. (1981). The energy supply model MESSAGE. IIASA, Laxenburg, Austria.

Tapetado, P. and Usaola, J. (2019). Capacity credits of wind and solar generation: The Spanish case. *Renewable Energy*, 143, 164–175.

Thiollière, N., Clavel, J.-B., Courtin, F., Doligez, X., Ernoult, M., Issoufou, Z., Krivtchik, G., Leniau, B., Mouginot, B., Bidaud, A. et al. (2018). A methodology for performing sensitivity analysis in dynamic fuel cycle simulation studies applied to a PWR fleet simulated with the CLASS tool. *EPJ Nuclear Science and Technology*, 4(13).

Tillement, S. and Hayes, J. (2019). Maintenance schedules as boundary objects for improved organizational reliability. *Cognition, Technology and Work*, 21, 497–515 [Online]. Available at: https://doi.org/10.1007/s10111-018-0530-y.

Toman, M. and Pezzey, J.C. (2002). The economics of sustainability: A review of journal articles. *Resources for the Future Discussion Papers*.

Turner, G.M. (2008). A comparison of The Limits to Growth with 30 years of reality. *Global Environmental Change*, 18(3), 397–411.

Van den Bergh, K. and Delarue, E. (2015). Cycling of conventional power plants: Technical limits and actual costs. *Energy Conversion and Management*, 97, 70–77.

Weimer-Jehle, W., Buchgeister, J., Hauser, W., Kosow, H., Naegler, T., Poganietz, W.-R., Pregger, T., Prehofer, S., von Recklinghausen, A., Schippl, J. et al. (2016). Context scenarios and their usage for the construction of socio-technical energy scenarios. *Energy*, 111(15), 956–970.

Wiesenthal, T., Dowling, P., Morbee, J., Thiel, C., Schade, B., Russ, H., Camara, G.D., Simoes, S., Peteves, E., Schoots, K. et al. (2012). *Technology Learning Curves for Energy Policy Support*. Publications Office of the European Union, Luxembourg.

Wilson, C., Grübler, A., Bento, N., Healey, S., De Stercke, S., Zimm, C. (2020). Granular technologies to accelerate decarbonization. *Science*, 368(6486), 36–39.

WNA (2020). Nuclear power in France.

World Energy Council (2019). Global energy scenarios comparison review.

Appendix

The Dismantling of Nuclear Facilities

Jacques PERCEBOIS[1] and Nicolas THIOLLIÈRE[2]

[1]*CREDEN, CNRS ART-Dev, Université de Montpellier, France*

[2]*Subatech, IMT Atlantique, Nantes, France*

Dismantling consists of deconstructing nuclear facilities (power plants, as well as research, fuel cycle, or waste management facilities). The goal is therefore to treat and package the waste, then store it if necessary, and store the waste by-products that are ultimately formed. In France, the medium and high radioactivity and long-life waste from the fuel are intended to be stored in a deep geological layer, as stipulated in the program law of June 28, 2006 on the sustainable management of radioactive materials and waste (see Chapter 1). Once the fuel has been extracted, there will be a large volume of waste generated from the dismantling, but in most cases this waste has low or very low radioactivity (FA and TFA). Figure A.1 shows the management channels implemented or planned according to the type of waste.

After the spent fuel has been removed from the power plant, the radioactivity present on the site is reduced by 99.9% with respect to the operating phase. The radioactivity remains present in the primary circuit, particularly in the reactor vessel. The deconstruction is then carried out in several successive steps:

– discharging spent fuel;

– emptying all circuits;

This appendix is written with the support of Sylvain Granger, director of deconstruction and waste projects at EDF.

– dismantling the primary circuit outside the tank;

– dismantling the tank;

– demolition of the buildings.

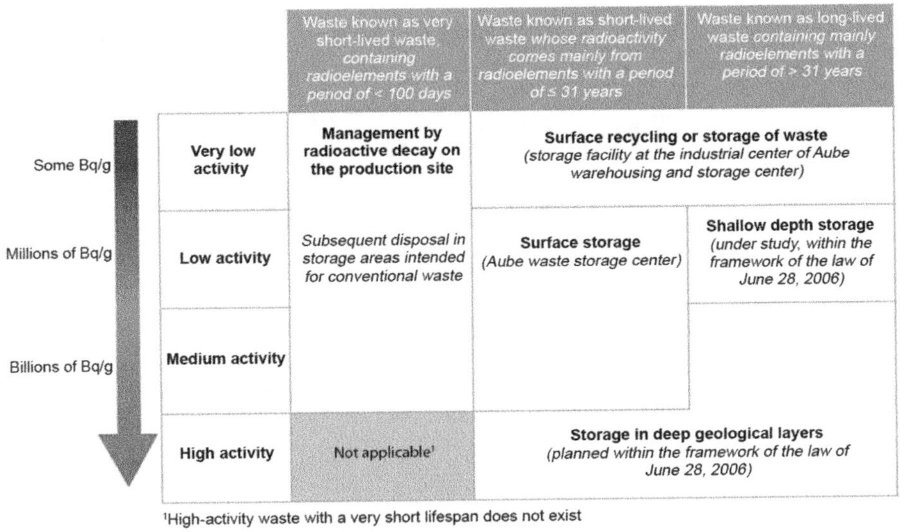

Figure A.1. *System for managing radioactive waste on the basis of category[1]. For a color version of this figure, see www.iste.co.uk/percebois/nuclear2.zip*

Note that France does not currently use a "release threshold", which means that even wastes whose radioactivity is significantly lower than the natural radioactivity remain radioactive wastes and cannot be managed in conventional waste streams or recovered for other uses. By comparison, Germany, like other European countries, recycles part of the metal components derived from the dismantling of its nuclear facilities, provided they are below the radioactive release thresholds defined in the European directive of 2013 (which guarantees a level of radioactivity that is insignificant compared to natural radioactivity). It should be noted that there is currently a project to revise regulatory texts in France with regard to this issue, which would make it possible to homogenize French practices with those of other European countries. However, the possibility of recovering metals resulting from dismantling will only be possible after a treatment is performed eliminating the radioactivity of these metals, and appropriate controls and verifications are undertaken by the independent control authorities.

1 Available at: https://www.ecologie.gouv.fr/demantelement-et-gestion-des-dechets-radioactifs.

In France, dismantling is declared by way of decree, establishing the description of the operations of the dismantling to be carried out, the primary stages, as well as the duration. To comply with this decree, a supporting file is created, which is then subject to an investigation by the Nuclear Safety Authority (*Autorité de sûreté nucléaire* [ASN]) and a public inquiry. The base nuclear installations (BNIs) can operate without fixed duration limits a priori under the supervision of the ASN. However, they will eventually be taken out of operation at some point, but their life does not end with the shutdown of the facility. In fact, it is necessary to proceed with the restoration of the site "in its initial state". The decree also sets the sanitation level of the site. At the end of the dismantling operations, the land that has been used for nuclear installations may be reusable for other purposes.

This section is limited to cases of dismantling nuclear power plants. A rigorous consideration should also be made of research facilities, fuel cycle facilities and waste storage sites. Interested readers may also consult (CEA 2017). Regarding nuclear installations dedicated to military applications, publicly available data are quite fragmentary and will not be discussed here.

A.1. The principles that govern dismantling in France

A.1.1. *The polluter-pays principle*

In France, it is the operators (EDF, as well as Orano, ex-Areva, and the CEA for their facilities) who have the technical and financial responsibility for the dismantling and the related radioactive waste, with no time limit; in some countries, it is the State that replaces the operator.

Thus, in Germany, the federal government is responsible for the surface storage and final storage of spent fuels and radioactive waste, but the dismantling of installations remains the responsibility of the operators. In the United Kingdom, dismantling is the responsibility of a state agency, the NDA (Nuclear Decommissioning Authority). After the definitive shutdown, the installation becomes the property of the NDA, which proceeds to dismantle it using subcontractors. Financing is done via the Nuclear Liabilities Fund, a fund supplied by the operators. In the United States, the operators are responsible for dismantling, but they have the option of transferring this responsibility to specialized private companies, subject the necessary funds being transferred to them at the same time. The long-term management of spent fuels is carried out by the federal government (Department of Energy) using funds generated by the operators during the production phase.

A.1.2. *The principle of an "immediate" dismantling*

Radioactivity decreases over time. This means it is tempting to wait to proceed with the dismantling, in order to allow access to the most radioactive areas, particularly the parts of the installation that were near the fuel supply during the operating phase. Thus, until the beginning of the 2000s, deferred dismantling was the rule at the international level, as it was within France. This approach nevertheless suffers from limitations, given that the level of radioactivity in certain areas can sometimes remain at a high level, even after several decades.

The principle of immediate dismantling was then established in France for several reasons. First of all, remote operation techniques have begun to reach an industrial scale that is sufficient to be used in hard-to-reach areas. The use of robots thus makes it possible to move faster in the dismantling process, since it is possible can access highly radioactive areas earlier. Next, the fact that France opened the Industrial Grouping, warehousing and Storage Center (in French: *Centre industriel de regroupement, d'entreposage et de stockage* [CIRES]) in Morvilliers in 2003 made it possible to provide a waste management chain resulting from the dismantling of nuclear power plants, consisting mostly of TFA waste. Finally, the French parliament rightly considers that it is up to the generations who have benefited from nuclear power to bear all the burdens it also brings, and not to leave it to future generations to carry out this deconstruction. The methods related to the dismantling of nuclear installations are defined within the framework of the law no. 2015-992 of August 17, 2015[2] on the energy transition for green growth. Article 593-25 provides that "when the operation of a basic nuclear installation or part of such an installation is definitively stopped, its operator will proceed to dismantle it as soon as possible, under economically acceptable conditions and in compliance with the principles set out in article L. 1333-1 of the Public Health Code and sec. II of article L. 110-1 of this Code".

Operators have the option to wait a few years before starting dismantling operations, but the length of this time period must remain reasonable. In practice, it is necessary to wait about 5 years before carrying out the actual dismantling work. This period is necessary to allow for the dismantling file to be investigated and to obtain the decree authorizing the operations. It is used to remove all the fuel, drain the circuits and adapt the facility to prepare for and facilitate dismantling operations.

The technology of the reactor has a major impact on the dismantling time. For a pressurized water reactor (PWR), such as those used in the network of EDF plants in operation and the two reactors at the Fessenheim site which were shut down in 2020,

2 Available at: https://www.legifrance.gouv.fr/loda/id/JORFTEXT000031044385/.

the breakdown of the duration of the dismantling depends on several factors (including the number of reactors per site), but we can consider it to be in the order of 10 years. This duration is longer for a graphite moderated reactor, such as the French UNGG or the British reactors, because the volume of equipment to be treated is much larger and the treatment of graphite presents additional complexities in comparison with metal and concrete components. The UNGG reactors thus contain about 20 times more equipment and materials than a REP by mass. Regarding FNR technology, the Superphénix reactor (1,240 MW), the largest reactor in the world under deconstruction, has a tank that is 24 m long, equivalent to 20 tanks from 1,300 MW REP reactors. At the same time, the treatment of the sodium-based coolant makes the operation much more complex, since it is necessary to take into account the chemical characteristics of this element.

Such experiences remain limited at the global level, although 169 reactors (mainly from the first and second generation) have already been completely shut down and their dismantling has already been initiated. In France, nine reactors are being deconstructed: the Brennilis heavy water reactor, the six UNGG reactors, the FNR at Creys-Malville (Superphénix) and the PWR at Chooz.

Chooz A is a central PWR of the same type as the network currently in operation in France, but with a power level of only 305 MW. This reactor was shut down in 1991. The complexity of dismantling this reactor can be explained by the fact that it is located within a cavern, which acts as a confinement enclosure. In time, this cavern will also be filled. A power level three times less than that of a standard 900 MW PWR does not automatically entail less equipment and electromechanical components to be disassembled and cut, and the removal of the components from the cavern presents particular difficulties (access problems). Authorized and initiated in 2007, the dismantling of the Chooz A reactor reached its final stage with the dismantling of the tank, which should be completed in 2022. The experiences obtained from Chooz A, which particularly demonstrated the effectiveness of cutting the tank underwater, was taken into account for the preparation of the dismantling of Fessenheim.

We note that dismantling two reactors on the same site will take less time than the dismantling of two reactors located on two different sites, simply because some equipment can be shared. There are no isolated reactors in France, but sites made up of two, four or six reactors, as the case may be. In addition, deconstruction operations are industrial activities with high added value, whose know-how is broadly capable of being exported.

In some countries, a "deferred" dismantling was the preferred choice from the outset, which means that a period of several decades was allowed to pass before carrying out the operations. Another option to consider is a "burial", which consists of pouring concrete on top of the facility on site, though this solution is not acceptable to the IAEA. In the United States, certain installations undergo immediate dismantling, and a few rare low-power installations have been placed under a sarcophagus, but the majority of facilities are subject to the practice of delayed dismantling (though this does not exceed 60 years after the final shutdown of the facility). The operator, who is required to respect a total duration of less than 60 years, can begin to dismantle part of the installation (the primary circuit in general), and then, after having reduced the risks very significantly, can suspend the operation for a few years before resuming it, with the goal of technical and financial optimization (seeking to find an optimum point between the reduction of risks and the use of funds).

A.1.3. *The principle of conservative cost estimation*

The insights gained from experiences (French: REX) are modest on a global scale with regard to dismantling. The so-called first-generation installations (the UNGG power plants, and Brennilis and Chooz A plants in France) have all been stopped, some have already been almost dismantled, and others are in the process of being dismantled. A graphite reactor costs more to dismantle than a pressurized water reactor due to the reasons mentioned above in the text.

Of the 342 second-generation PWR-type reactors commissioned worldwide in the 1970s, 1980s, and 1990s, 50 have been definitively shut down; only seven facilities have been totally dismantled (all of which are located in the United States), while the others have been dismantled according to the IRSN (2020).

The costs of deconstruction appear to vary widely. These costs vary from the initial level, depending on the reactor technologies and the series effects from which benefits can be drawn. Given the previous experiences accumulated over time, EDF estimates that the cost of dismantling an isolated second-generation PWR reactor (without the final waste management costs) is in the order of 500 million euros (EDF 2019).

For a network of PWR reactors, the average cost can be significantly reduced thanks to pooling effects (where several reactors are dismantled on the same site) and serial effects (the reuse of parts of studies and tool sets between several operations). These effects make it possible to estimate an average cost in the order of 350 million euros per reactor for the network in France. With regard to third-generation EPRs, their dismantling costs will probably be higher, given that their construction uses more concrete and reinforcement elements.

A.1.4. *The principle of pre-financing*

Operators must make provisions on the basis of anticipated costs, and in France, these provisions must be secured on the assets of the balance sheet, in the form of dedicated funds. For EDF, these consist of investments outside the nuclear industry, in the form of shares, bonds, real estate and infrastructure assets. For example, the part of EDF's provisions is constituted by 50% of the value of the RTE, the electricity transmission network of which EDF is the majority shareholder. A commission, the CNEF[3], responsible for examining reports for Parliament, ensures in principle that these provisions are genuine and secure, but its role appears to be limited. In practice, it is the DGEC (Ministry of Energy) which is the authority in charge of ensuring the security of the financing for future dismantling and waste management expenses. In particular, it monitors the evaluations of the provisions made by the operators and that the management of the dedicated assets is done properly. Nuclear operators publish a report on these topics every 3 years with annual updates. These reports, which are also made public, are examined by the DGEC, which may request additional information, call for modifications, or carry out external audits by specialized firms.

In this way, in 2016, an independent external audit commissioned by the DGEC concluded that EDF's provisions for the future dismantling of the operating reactor fleet were prudent. It should be noted that in some countries, the operator pays into an external fund that, at the appropriate time, will be in charge of financing the operations. This fund is managed by the state or by an ad hoc organization. The fact that the provisions are constituted internally in the case of France reflects parliament's desire to make the operator responsible while securing future financing even in the event that the operator declares bankruptcy, given that since these assets are dedicated, they cannot be used for other purposes and would be transferred automatically to the government, which would take control. Note that these assets generate interests, and that such interests are often higher than the discount rate used for the management of provisions. The nominal discount rate generally used in France is 4.4%, while the return on assets observed over the last 10 years is around 6.5%. The maximum value of the discount rate is set by the state and reviewed regularly. In Germany, operators have also placed provisions in the liabilities of their balance sheets, but unlike in France, these provisions are not allocated to

3 National Commission for the evaluation of the financing of the costs for the dismantling of base nuclear installations and for the management of spent fuels and radioactive waste.

dedicated financial assets, which presents a dual risk to liquidity and the coverage of future expenses.

The diagram shown in Figure A.2 presents the chronology of the operations. It should be noted that the installed power that must be dismantled is much higher in the second generation than in the first generation.

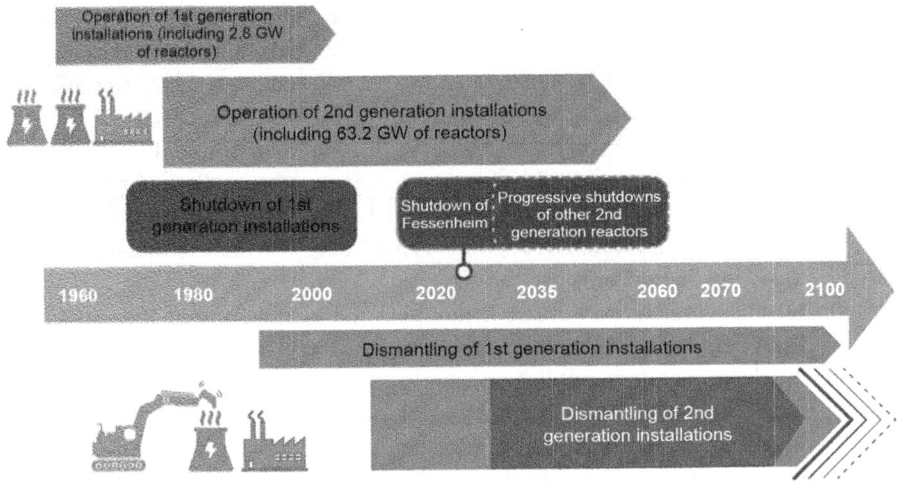

1st generation facilities: **about twenty reactors**: approximately 10 experimental and research reactors (CEA), 6 UNGG reactors (EDF), 1 heavy water reactor (EDF), 1 REP prototype reactor (EDF), reprocessing plants UP1 (CEA) and UP2-400 (Orano); Marcoule facilities (workshops, laboratories, and treatment stations) and other CEA facilities (STED laboratories in Grenoble, etc.).

2nd generation installations: 2 fast neutron reactors (Phenix-CEA and Superphenix EDF), **58 pressurized water reactors (REP)** currently in operation (EDF), reprocessing plants UP2-800 and UP3 (Orano), Georges Besse enrichment plants (Orano), etc.

Figure A.2. *Chronology of the lifecycle of nuclear installations (source: Cour des comptes 2020). For a color version of this figure, see www.iste.co.uk/percebois/nuclear2.zip*

A.2. Dismantling costs

The waste created from dismantling works (excluding spent fuel) is 80% conventional waste, consisting of rubble, concrete, scrap metal, other metals and 20% radioactive waste. The net charges (and therefore not updated) for the dismantling of the nuclear plant network in France (reactors from the first generation and the 58 PWRs from the second generation) were estimated at the end of 2018 at 46.373 billion euros at 2018 value, including 27.331 billion for EDF, 9.128 billion for Orano, and 9.914 billion for the CEA (Cour des comptes 2020). These figures do not include the costs of storing radioactive waste (see Chapter 1), which depending on the case are estimated at between 25 and 35 billion euros.

Given in euros per kW, the figure in France is significantly lower than the global average: in 2016, the cost for dismantling had been estimated at 450 $euros_{2013}$/kW in the United States on average, versus 306 $euros_{2013}$/kW in France, according to the Court of Auditors (2020 report). The planned standardization of operations could explain this difference, since the second-generation reactors in France are all of the same type. Moreover, it is now recognized that comparing cost amounts on the basis of installed capacity does not really make sense, since as we have seen previously, these costs can vary greatly depending on the reactor technologies, and within the same technology, depending on the level of pooling and serial effects, and the quantity of equipment to be dismantled (see the example of Chooz A). The availability and efficiency of waste management systems is also an important factor that can have significant consequences on the feasibility, planning and cost of operations. France is currently one of the few countries to have complete industrial sectors for managing radioactive waste.

In general, according to the report (Berger and Assystem 2020), the studies show that the cost of dismantling a reactor with a power level of 900–1,000 MW varies greatly from one reactor to another worldwide, with the average being within the range of 500–550 million euros (i.e. within the range of slightly higher than 550 euros per kW). The producer EDF is instead placed in the lower part of the investigation, with a cost between 300 and 350 million euros per PWR reactor, due to the pooling and serial effects estimated at the network level.

To confirm the level of 500 million euros for an isolated PWR (within one operation on a single reactor within a site), one interesting case to examine is the situation in the United States for five PWRs whose dismantling has now been completely finished. A diverse array of costs can be seen clearly in Table A.1, which can be explained by the high contrast between situations that in some cases are not comparable to the cases in France. However, the two reactors most comparable to the French case (excluding pooling and serial effects), Maine Yankee and Rancho Seco, confirm the precedents set for general cost levels. Therefore, better feedback and economies of scale should make it possible to both reduce certain costs and to limit the standard deviation that will be observed in the future.

	Type	Power (MW)	Cost (M$_{2016}$)	Cost (M€_{2016})
Connecticut Yankee	PWR	560	931	887
Maine Yankee	PWR	860	500	476
Rancho Seco	PWR	913	518	493
Trojan	PWR	1,180	300	286
Yankee Rowe	PWR	167	750	714

Table A.1. *Examples of finished dismantling operations and their associated costs (data from Aubert and Romagnan 2017)*

A.3. Financing for dismantling

The funds provided must make it possible to cover the dismantling costs at the proper time. These provisions consist partly of government bonds (OAT), partly of private bonds, partly of company stock and partly of investments in real estate and infrastructure. The rest is made up of liquid assets.

It is not necessary for the provisions given now to be equal to the amount of expenses that will have to be covered tomorrow. The financial assets that are set aside will in fact generate income in the form of interest or dividends, and these types of income will be added to the initial capital. In other words, the gross expenses by their nature are higher than the provisions recorded in the company's accounts due to the discounting logic, as illustrated in Figure A.3.

The fact that the projected gross dismantling expenses are due only at maturity, at the time when this dismantling will begin, gives justification for the fact that only a fraction of the overall envelope of gross expenses is provisioned. Note in this regard that an extension of the life of the current network is likely to make things easier for the operator, since the accumulated provisions will generate more income in the future.

At the end of 2018, provisions for the dismantling of power plants in France totaled to 28.257 billion euros (compared to the gross expenses of 46.343 billion euros). These provisions were then distributed among the three operators: 15.985 billion euros for EDF, 5.060 billion for Orano and 7.212 billion for the CEA. These provisions do not take into account the so-called "last hearts" cost (estimated at about 4 billion and provisioned elsewhere). Thus, the amount of provisions, at 28.257 billion euros, should cover the 46.343 billion euros of expenses in gross value at the appropriate time and based on the discount rate that is set.

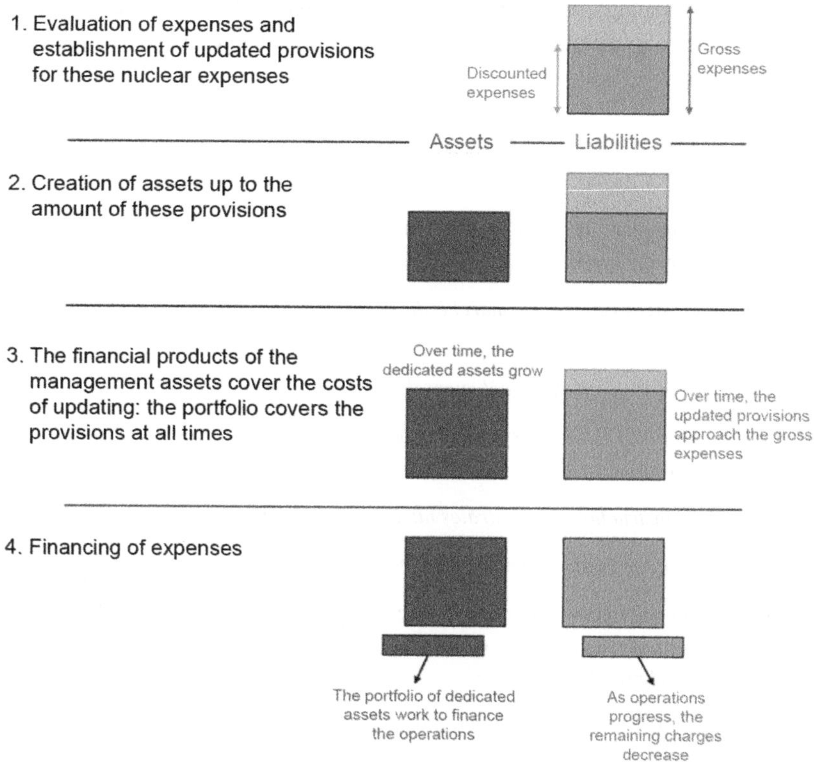

Figure A.3. *Graphic representation of the mechanism used for financing dismantling expenses (source: Ministry of Ecological Transition 2020). For a color version of this figure, see www.iste.co.uk/percebois/nuclear2.zip*

For its part, the choice of the discount rate is important. The current rate (1.1%) is chosen by the operator, and it must remain below a maximum rate set by the DGEC, which is currently 4.5%. This rate is controlled independently and is reviewed annually. It must be lower than the rate of return on dedicated assets. As indicated above, the insights gained from the experiences in France show that this has been the case for 15 years, since over this period the rate of return on assets has been close to 6%.

It is still necessary for these provisions to be secured, which is done through the formation of dedicated assets. It is clear that in the event the operator is dissolved, the State will take over, but it will also have access to the necessary funds. In addition to this dismantling cost, we will of course also need to add the costs of radioactive waste management, which are also provisioned and secured by dedicated assets. This point is discussed in another section of the work (see Chapter 1 section 1.2.3).

A.4. References

Aubert, J. and Romagnan, B. (2017). Rapport d'information sur la faisabilité technique et financière du démantèlement des INB.

Berger, R. and Assystem (2020). *Le marché du démantèlement nucléaire en Europe*. Munich, Germany.

CEA (2017). *L'assainissement-démantèlement des installations nucléaires*. CEA.

Cour des comptes (2020). L'arrêt et le démantèlement des installations nucléaires. Cour des comptes, Paris, France.

EDF (2019). *La déconstruction des centrales nucléaires*. EDF, Paris, France.

IRSN (2020). Les démantèlements des réacteurs nucléaires dans le monde. IRSN, Paris, France.

Ministère de la Transition écologique et solidaire (2020). Démantèlement et gestion des déchets radioactifs ; coût et financement des charges à long terme. Ministère de la Transition écologique et solidaire, Paris, France.

OECD/NEA (2019). L'énergie nucléaire en 2019. OECD/NEA, Paris, France.

Quinet, A. (2019). Quelle valeur donner à l'action pour le climat ? *Économie et Statistiques*, 510(12), 171–187.

List of Authors

Jacques CANTARELLA
ONDRAF
Brussels
Belgium

Patrick CRIQUI
CNRS
Université Grenoble Alpes
France

Cécile EVANS
Orano
Châtillon
France

Pierre KUNSCH
ONDRAF
and
École polytechnique de Bruxelles
Brussels
Belgium

Lionel LEMIALE
IAE Nantes
France

Didier LÉONARD
ONDRAF
Brussels
Belgium

Rodica LOISEL
IAE Nantes
France

Jean-Paul MINON
ONDRAF
Brussels
Belgium

Ludivine PASCUCCI-CAHEN
Laboratoire d'analyse économique des risques nucléaires
IRSN
Fontenay-aux-Roses
France

Jacques PERCEBOIS
CREDEN, CNRS ART-Dev
Université de Montpellier
France

Boris SOLIER
CREDEN, CNRS ART-Dev
Université de Montpellier
France

Nicolas THIOLLIÈRE
Subatech
IMT Atlantique
Nantes
France

Xavier URSAT
Groupement des industriels français de l'énergie nucléaire
Paris
France

Index

Summary of Volume 1

Preface

Sylvain DAVID, Jacques PERCEBOIS and Nicolas THIOLLIÈRE

Chapter 1. The Evolution of Nuclear Energy in the World and in France

Daniel IRACANE, Stéphanie TILLEMENT and Frédéric GARCIAS

Printed and bound by CPI Group (UK) Ltd, Croydon, CR0 4YY

13/09/2026

14960997-0004